CAMBODIA AGRICULTURE, NATURAL RESOURCES, AND RURAL DEVELOPMENT SECTOR ASSESSMENT, STRATEGY, AND ROAD MAP

JULY 2021

ASIAN DEVELOPMENT BANK

ADB

© 2021 Asian Development Bank
6 ADB Avenue, Mandaluyong City, 1550 Metro Manila, Philippines
Tel +63 2 8632 4444; Fax +63 2 8636 2444
www.adb.org

Some rights reserved. Published in 2021.

ISBN 978-92-9262-951-9 (print); 978-92-9262-952-6 (electronic); 978-92-9262-953-3 (ebook)
Publication Stock No. TCS210256-2
DOI: http://dx.doi.org/10.22617/TCS210256-2

Notes:
In this publication, "$" refers to United States dollars.
ADB recognizes "China" as the People's Republic of China, "Laos" as the Lao People's Democratic Republic, "South Korea" as the Republic of Korea, and "Vietnam" as Viet Nam.

On the cover: Enhancing sustainable agricultural productivity and strengthening agricultural value chains are among the key areas of support by ADB in Cambodia's agriculture sector (photo by ADB).
Cover design by Josef Ilumin.

Contents

Tables and Figures

Tables

Figures

Abbreviations

ADB	Asian Development Bank
ANRRD	agriculture, natural resources, and rural development
COVID-19	coronavirus disease
DRR	disaster risk reduction
FAO	Food and Agriculture Organization of the United Nations
FiA	Fisheries Administration (MAFF, Cambodia)
FWUC	farmer water user community
GDP	gross domestic product
GMS	Greater Mekong Subregion
ICT	information and communication technology
IFAD	International Fund for Agricultural Development
MAFF	Ministry of Agriculture, Forestry and Fisheries
MFI	microfinance institution
MOWRAM	Ministry of Water Resources and Meteorology
NGO	nongovernment organization
O&M	operation and maintenance
PRC	People's Republic of China
SMEs	small and medium-sized enterprises
SNAP-DRR	Strategic National Action Plan for Disaster Risk Reduction

Weights and Measures

ha	hectare
kg	kilogram
km	kilometer
mcm	million cubic meter

Executive Summary

The sector assessment, strategy, and road map describe the performance and development constraints of Cambodia's agriculture, natural resources, and rural development (ANRRD) sector in light of the coronavirus disease (COVID-19) pandemic and the current strategic investment priorities of the Government of Cambodia and the Asian Development Bank (ADB).

Sector Overview and Assessment

Cambodia became a lower middle-income economy in 2015, after sustaining an average annual gross domestic product (GDP) growth rate of 7.6% from 1994 to 2015. In 2016–2019, total GDP grew by 7.1% annually and contracted by 3.1% in 2020 due to the negative impact of the COVID-19 pandemic. Garment exports, agriculture, tourism, and more recently construction and real estate have been the drivers of economic growth since 2000.

The share of agriculture value-added to the total GDP averaged around 30% during 2000–2014 and was estimated at 22.1% in 2019. Within the sector, crop production has the largest output contribution to total national GDP at 13.0%, followed by fisheries (5.5%), livestock (2.6%), and forestry (1.6%). However, the growth of agriculture value-added has been limited (0.85% annually, on average) since 2013 due to the long-term low labor productivity stemming from the preponderance of unprocessed rice production and the downturn in international commodity prices since 2014, as well as adverse weather conditions in 2015 and 2016.

Impact of the COVID-19 pandemic. For the global food trade, COVID-19 has impacted both supply and demand sides, depending on the severity of the spread of the disease. During the pandemic, the Cambodian government has allowed the transport of goods nationwide, including agricultural inputs and products. As a result, prices have remained fairly stable for most key food commodities over the first 10 months of 2020, except for an initial spike in prices at the end of March 2020 and the beginning of April 2020. The impact of the pandemic is considered on the demand side, with many households suffering loss of livelihoods and income as a result of the pandemic, which in turn has restricted their ability to afford sufficient and nutritious foods. In response, the government launched a cash transfer program in June 2020 to provide support to vulnerable households, reaching around 696,000 poor households by October 2020.

Agriculture

Limited postharvest handling and processing. Only about 10% of Cambodia's total agricultural outputs are processed within the country, whereas processed agricultural exports represent only 8% of total official exports by value. Other than rice milling (with its expanded capacity) and basic processing of rubber and pepper, the agro-processing sector is largely underdeveloped. Cambodia needs private investment in postharvest handling and processing if it is to capture economic opportunities arising from value addition to its agricultural products.

Key constraints to increased agro-processing in Cambodia include the reliability and cost of electricity, informal payments for permits and documentation, lack of skills and information, difficulties in obtaining development capital, and access to technology and machinery.

Access to agricultural inputs and machinery. Modernizing agriculture and increasing labor productivity for farm production require quality agricultural inputs and mechanization. Cambodia has limited availability of improved seed varieties. Although the annual supply of quality rice seeds has been increasing, the supply was only 20% of the annual demand in 2019. Vegetable seeds are imported except for those produced at government research stations. The Government of Cambodia has not yet ratified a seed policy with seed quality standards under the 2008 Seed Law. Moreover, it has yet to make the seed law operational by establishing the essential policies and enforcement mechanisms. Cambodia's fertilizer use per hectare of cropland increased from 10.0 kilograms in 2005 to 33.0 kilograms in 2018. Pesticide use has increased, especially during dry-season rice cultivation and in the vegetable sector. Pesticides are also imported; however, most pesticides are allegedly either poorly regulated or unregulated. Proper use of regulated pesticides is important for farmers' safety, food safety, and ecosystem health. Cambodian agriculture has seen a gradual mechanization and replacement of labor with machinery such as power tillers and water pumps. Further mechanization is expected to increase labor productivity and earnings from farming to match income opportunities outside farming.

Access to markets, transport, and logistics. Farmers are predominantly price takers. Most sell their crops to traders just after harvest. Varieties are mixed and only small volumes of pure varieties are found, which hinder value recognition and addition. For further value addition in agriculture, access to international markets is increasingly important. Efficient transport and logistic infrastructure and services are necessary to strengthen agri-food value chains. Roads are the major means of domestic transport in Cambodia. Cambodia's domestic transport costs have been reduced, but they remain higher than neighboring countries such as Thailand and Viet Nam. Road improvements and increased competition among local transport services are needed to reduce transport costs. Similarly, Cambodia's physical market infrastructure is outdated, congested, unsanitary, and requires significant improvement. There is also no reliable cold chain system in place to ensure the proper handling and safe storage and distribution of perishable agricultural and food products.

Access to information and communication technology. The government has recently undertaken several policy interventions toward digital development, including its Information and Communication Technology (ICT) Masterplan 2020 and the draft Cambodia e-Government Master Plan, 2017–2022. However, the country still ranks low on digital adoption and technological readiness. ICT can play an important role in linking agriculture value chain stakeholders such as farmers, agribusiness operators, consumers, and government agencies. The digitization of information will facilitate the formation of value chain platforms through which value chain stakeholders can exchange information, services, and products.

Access to finance. Despite the increased access to finance in rural areas, credit to agriculture and agribusinesses together accounted for only 9.4% of formal finance in Cambodia in 2018, and access to finance is still the major bottleneck for Cambodian farmers and agribusinesses. High collateral requirements, high interest short-term loans, lack of credit history, and low financial literacy are among the reasons for their poor financial access. Furthermore, there is some indication that, due to the COVID-19 pandemic, access to credit and finance has been restricted. Addressing constrained access to finance will be critical to spur post-pandemic recovery.

Extension services and cooperatives. Although about 70% of all villages are covered by some type of public or private extension services, they remain generic and tend to be supply-driven. The government is in the process of reorienting its extension services to be demand-driven, decentralized to enhance outreach, and varied to support agriculture diversification. While improving the government extension services, engagement of private service providers is important, as is close coordination with extension services through donor-funded projects. The government promotes farmer cooperatives to take advantage of economies of scale in production and marketing arrangements. There were around 1,200 registered cooperatives in 2019; however, their capacity varies. Further strengthening of farmer cooperatives and different forms of farmer groups is required for sustainable farming sector development in the country.

Natural Resources

Water resources. Cambodia experiences seasonal water scarcity resulting primarily from lack of stored water and limited access to water and not from the volume of water. In agriculture, increasing national water use productivity entails better water harvesting and storage capacity on farms, and improved water use efficiency (e.g., through the use of drip irrigation systems). However, the country's relatively flat topography offers limited potential for water harvesting and storage facilities to improve water availability. Currently, only 22% of the country's 4.5 million hectares of arable land (including 3.2 million hectares of cultivated rice area) are covered by 2,480 irrigation schemes. However, most are not fully functioning due to poor operation and maintenance, thereby constraining agricultural productivity. A consensus among development partners is that there is a need for a shift in focus from investment in irrigation infrastructure to a more modern and wider water resources management strategy geared toward increased agricultural productivity while sustaining key ecological processes.

Forests and biodiversity. A variety of forest types covers almost half the country's land area. Cambodia's forests and its coastal and riverine ecosystems provide habitats for diverse species of plants and wildlife, including around 6,500 native flora and fauna species, some of which are endangered. The country's forests serve as ecological buffers to natural disasters and protect watersheds, act as carbon sinks, reduce soil erosion thereby averting fertility loss, and prevent flooding. Cambodia's forest cover has undergone a continuous and dramatic decline from 60% of total land area in 2006 to nearly 50% in 2018. Much of the remaining 9.0 million hectares of forest cover have been degraded as a result of selective logging, unregulated fuelwood extraction, indiscriminate awarding of economic land concessions, and other unsustainable uses of forest resources for national development priorities. Toward Cambodia's Sustainable Development Goal target of 50% of national area under forest cover, the government has adopted a conservation corridor approach, which needs to be supported with adequate investment.

Crosscutting Issues

Food security, safety, and nutrition. In recent years, the food security situation in Cambodia has improved considerably due to the growth of real income and increases in the production volume of rice and other crops. However, the COVID-19 pandemic threatens to reverse this trend through losses of income opportunities due to the economic downturn. Malnutrition remains prevalent among the poorest and most vulnerable sectors (i.e., the landless, female-headed households, the disabled, ethnic minorities, and those living in the most remote and marginalized areas). Food safety is also a concern with the pending passage of the Food Safety Law, a lack of policy and technical standards on food safety control and management, inadequate coordination of concerned government agencies, and poor enforcement of food safety regulations.

Climate change and disaster risk management. According to several risk indexes, Cambodia is rated *highly vulnerable* to the impacts of climate change (an increase in precipitation, temperatures, and intensity and frequency of climate hazards). Without adaptation to these climate effects, Cambodia is projected to lose 10% of GDP by 2050. The agriculture sector is particularly vulnerable to climate-related disasters. Floods and droughts can seriously hamper agricultural productivity and damage crops extensively. Pests and diseases such as avian influenza and swine flu can reduce food production and bring economic losses to farmers. Institutionalizing disaster risk reduction is crucial to strengthening Cambodia's technical capacity for disaster risk management, assessment, and monitoring.

Sector Strategy

Key areas of support. ADB's ANRRD strategy and program are in line with the Cambodian government's sector strategic development plans and objectives, and prioritize three key areas of support: (i) enhancing agricultural productivity, including improving land use and water resources management through a whole-of-system approach, improving irrigation efficiency and sustainability, investing in pre- and postharvest facilities and technologies, and making on-farm practices more efficient and sustainable; (ii) strengthening agricultural value chains, including promoting commercialization and market connectivity, increasing farm income through crop diversification as well as value addition and mechanization, strengthening agricultural cooperatives and developing agribusiness networks, and encouraging private sector engagement; and (iii) strengthening natural resources management and disaster resilience, and mainstreaming climate change. These three areas complement each other to support the ANRRD sector's transformation toward a more productive, value-additive, and resource-efficient sector.

Crosscutting themes, risks, and assumptions. ADB's strategy and program will incorporate the following crosscutting themes: (i) governance and capacity building, (ii) cross-sector synergies, (iii) regional cooperation and integration, and (iv) gender development. The strategy is based on the assumption of continued expansion of subregional and regional trade opportunities, notwithstanding the not-so-optimistic world commodity outlook and the possible reorientation of Cambodia's export strategy to high-growth markets. Likewise, the strategy faces risks from (i) weak institutional and legislation commitment to reforms in relation to government policies and plans; (ii) uncertainties and knowledge gaps with respect to some untested technologies and practices (e.g., climate-smart agriculture); (iii) unclear operational provisions of some elements of government policy (e.g., approval and management of public–private partnership arrangements); (iv) prevailing conditions when finance is made available to the ANRRD small and medium-sized enterprises; (v) government's absorptive capacity; and (vi) impacts of the COVID-19 pandemic on the economy and the ANRRD sector, specifically on food security and value chains.

I. Context and Strategic Issues in the Sector Assessment

This sector assessment, strategy, and road map report describes the current strategic investment priorities of the Government of Cambodia and the Asian Development Bank (ADB) in Cambodia's agriculture, natural resources, and rural development (ANRRD) sector.

Rice field in Cambodia. Rice continues to be Cambodia's major crop, principal food, and most important export commodity (photo by Asian Development Bank).

Country Context and Overview of the Agriculture, Natural Resources, and Rural Development Sector

Recent Macroeconomic Trends

High economic growth for 2 decades. Cambodia sustained an average annual growth rate of 7.6% from 1994 to 2015 and became a lower-middle-income economy in 2015.[1] Real national income more than tripled over these 2 decades. In 2019, gross domestic product (GDP) per capita was reported at $1,696.[2] This high economic growth has been driven by garment exports, agriculture, tourism, and, more recently, construction and real estate. Figure 1 depicts the yearly growth rate of total GDP and agriculture, forestry, and fishing value-added growth. Total GDP in 2016–2019 grew at 7.1% per annum, on average. However, the economy contracted by 3.1% in 2020 because of

---

[1] ADB. 2019. *Key Indicators for Asia and the Pacific 2019*. Manila. In 1994–2015, Cambodia placed sixth among the world's fastest-growing economies.

[2] Government of Cambodia, Ministry of Planning, National Institute of Statistics. 2020. *National Accounts Statistics 2020*. Phnom Penh.

Figure 1: Growth in Gross Domestic Product (Total vs. Agriculture), 2000–2020
(annual % growth)

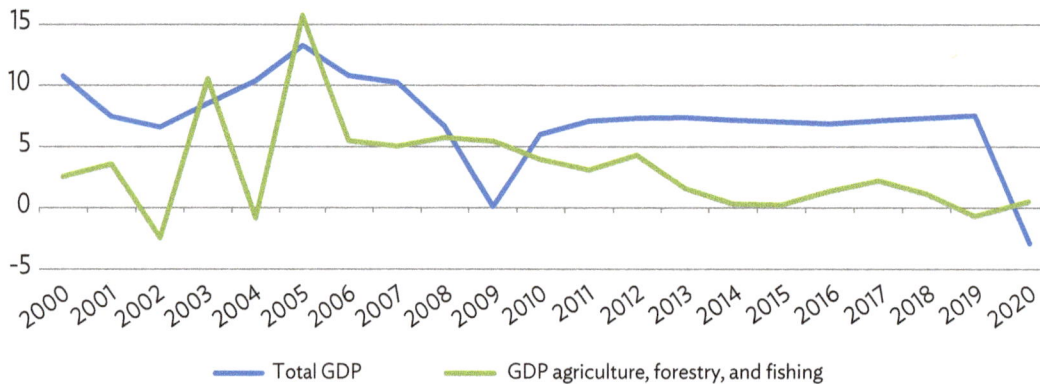

GDP = gross domestic product.

Sources: Asian Development Bank. 2021. *Asian Development Outlook 2021: Financing a Green and Inclusive Recovery*. Manila; and World Bank national accounts data, and Organisation for Economic Co-operation and Development National Accounts data files (accessed 7 June 2021).

the negative impact of the coronavirus disease (COVID-19) pandemic.[3] It is forecast to rebound, with projected growth of 4.0% in 2021 and 5.5% in 2022, as the economic recovery in major trading partners spurs demand for Cambodia's exports (footnote 3). The relatively high economic growth prior to the pandemic was built on the country's openness to trade and capital flows and the stable exchange rate policy against the United States dollar. Additional drivers included preferential access to key export markets, the receipt of large amounts of official development assistance, and even larger foreign direct investment inflows.[4] Agriculture production grew marginally by 0.5% in 2020 but is expected to increase by 1.3% in 2021 and 1.2% in 2022, bolstered by improved crop production after the flood damage in October 2020, sustained growth in aquaculture, and higher agriculture exports to the People's Republic of China (PRC) from a new bilateral free trade agreement (footnote 3). The country had a relatively low inflation rate, at 1.9% in 2019, which increased to 2.9% in 2020 and 3.1% projected in 2021 as a result of rising food and fuel prices (footnote 3).

Agriculture in the Economy

High share of agriculture in gross domestic product and employment. In the 1990s, the value-added share of agriculture in the total GDP fell from 35.7% in 2000 to 22.1% in 2019 (Figure 2).[5] Cambodia's GDP share of agriculture is the highest among the lower-middle-income countries.[6] In terms of the agriculture sector's

---

[3]   ADB. 2021. *Asian Development Outlook 2021: Financing a Green and Inclusive Recovery*. Manila. For additional macroeconomic indicators, refer to ADB's dataset on Key Indicators for Cambodia.

[4]   Foreign direct investment (FDI) net inflows in 2019 were 13.2% of GDP (at $3,561 million, almost the highest in the world in proportion to GDP). The share of FDI inflows to the agriculture sector has been shrinking from 13.6% of total net inflows in 2015 to 5.8% in 2019. Net official development assistance receipts are in decline, from 12.2% of GDP in 1996 to 5.0% of GDP in 2014, and less thereafter.

[5]   ADB. 2020. *Key Indicators for Asia and the Pacific 2020*. Manila. Estimates of GDP share vary according to whether constant or current prices are used. The estimate of the Ministry of Agriculture, Forestry and Fisheries (MAFF) is 26.3% from its 2016–2017 Annual Report and 24.9% from its 2017–2018 Annual Report.

[6]   The average share for lower-middle-income countries, including Bangladesh, the Philippines, and Viet Nam, is 18.0%.

Figure 2: Total Gross Domestic Product and Share of Agriculture Value Added, 2000–2019

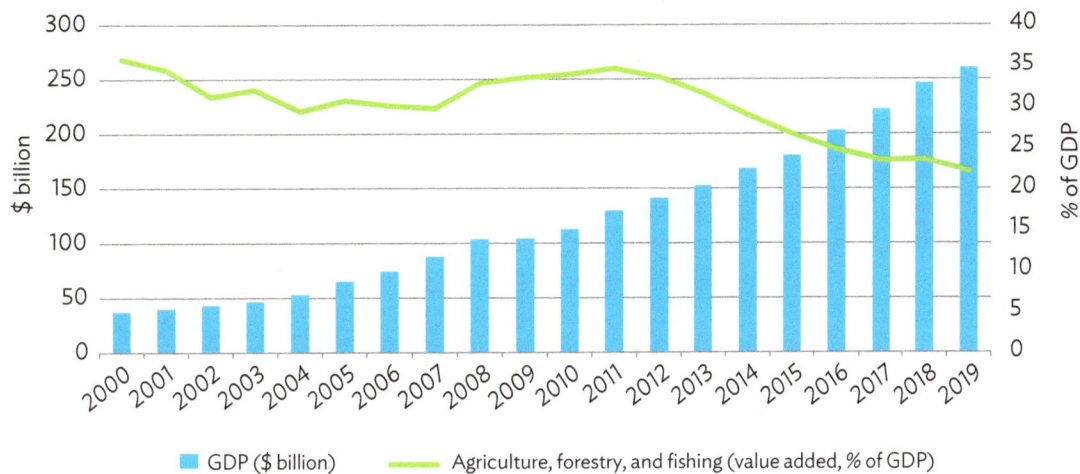

GDP = gross domestic product.

Source: World Bank national accounts data, and Organisation for Economic and Co-operation and Development National Accounts data files (accessed 7 June 2021).

output contribution to total national GDP, crop production still leads (13.0%), followed by fisheries (5.5%), livestock (2.6%), and forestry (1.6%).[7] Agriculture sector growth has been limited annually to 0.85%, on average, since 2013. This was caused by the long-term low labor productivity stemming from the preponderance of unprocessed rice production and the downturn in international commodity prices since 2014, as well as adverse weather (rain and drought) conditions in 2015 and 2016.[8] The proportion of the rural population is 76.2% of the national population, which is higher than the average of lower-middle-income countries (at around 60%).[9] In 2019, it was estimated that 2.99 million people were engaged in the agriculture sector (equivalent to 32.3% of the total labor force).[10]

Impact of the COVID-19 pandemic. The annual GDP growth rate is expected to decline dramatically to –3.1% in 2020. For the global food trade, the pandemic has impacted both supply and demand sides, depending largely on the severity of the spread of the disease. Several Southeast Asian countries temporarily imposed border controls, which disrupted the flow and prices of commodities, particularly in the second quarter of 2020. Cambodia imposed a ban on rice exports in April 2020 but lifted it in May 2020. Due to the short duration of the ban and the good harvest in 2020, Cambodia is expected to export 1.5 million tons of rice, 5.4% above the 2019 level.[11] During the pandemic, the government has allowed the transport of goods nationwide, including agricultural inputs and products. As a result, prices remained fairly stable for most key food commodities over the

[7]   Government of Cambodia, MAFF. 2019. *Annual Report, 2018–2019*. Phnom Penh.

[8]   World Food Programme, United Nations Children's Fund (UNICEF), and Food and Agriculture Organization of the United Nations (FAO). 2016. *Household Resilience in Cambodia: A Review of Livelihoods, Food Security and Health*. Part 1: 2015/2016 El Niño Situation Analysis. Phnom Penh.

[9]   International Labour Organization (ILO). ILOSTAT Database. Population by rural/urban areas. UN estimate and projections. July 2019.

[10]   ILO. ILOSTAT Database. Employment by sex and economic activity. ILO modeled estimates. November 2019.

[11]   FAO. Global Information and Early Warning System (GIEWS). Country Briefs: Cambodia (accessed 5 January 2021).

first 10 months of 2020, except for an initial spike in prices at the end of March and the beginning of April.[12] The impact of the pandemic is considered on the demand side, with many households suffering the loss of livelihoods and income, which in turn has restricted their ability to afford sufficient and nutritious foods. In response, the government launched a cash transfer program in June 2020 to provide support to vulnerable households, reaching 696,000 poor households by October 2020 (footnote 12).

Limited value-added of agro-processing. An increase in value addition to agriculture is crucial to increasing rural employment in the economy. However, national GDP figures for 2019 suggest that the combined total manufacturing value of agricultural-based products is only 3.4% of GDP.[13] This is due to the low level of private sector investment. Large-scale investments in agriculture and agro-industry were less than 10% of all investments, with investment in food processing being less than 2% of all investments.[14]

The majority of the country's agriculture exports are still in raw form and heavily dependent on crops. Figure 3 shows that the total value of agricultural exports has more than tripled since 2010. The main export products are rice and rubber, with an increasing share of cassava. From 2010 to 2018, the biggest reduction of export value was seen in maize.

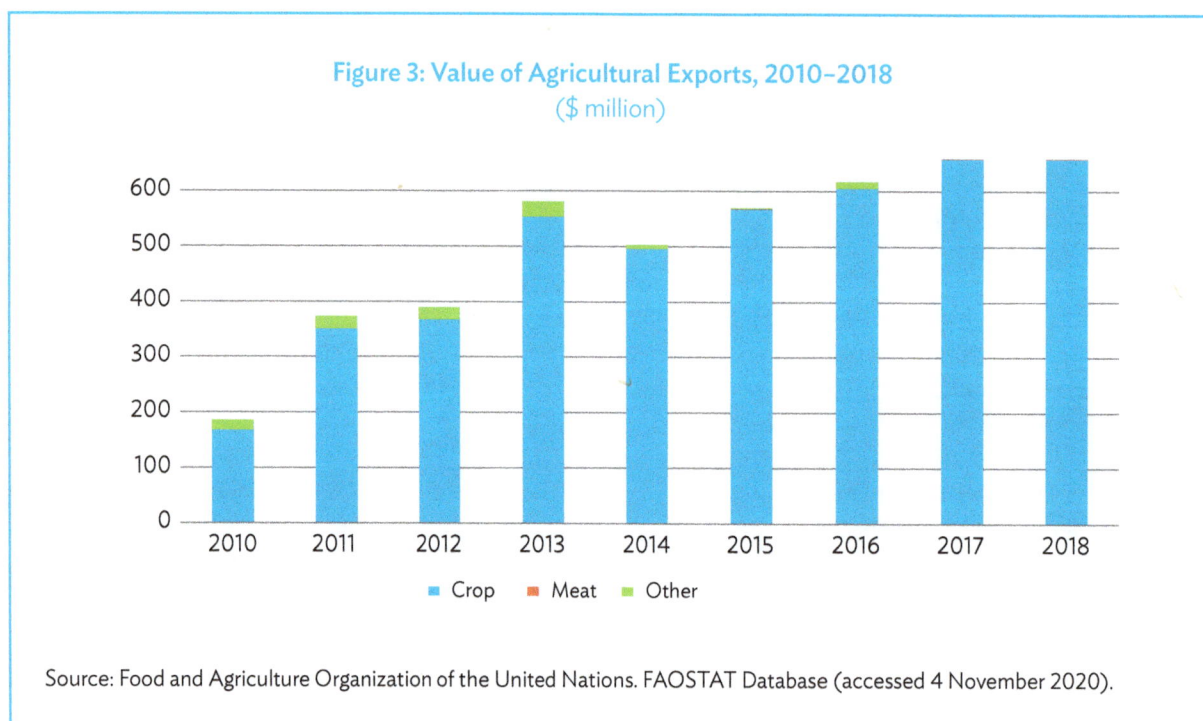

Figure 3: Value of Agricultural Exports, 2010–2018
($ million)

Source: Food and Agriculture Organization of the United Nations. FAOSTAT Database (accessed 4 November 2020).

---

[12] Vulnerability Analysis and Mapping Unit, Cambodia Country Office, World Food Programme. 2020. *Focus: Effect of the COVID-19 Outbreak on Food Prices.* PowerPoint presentation prepared for the World Food Programme on Cambodia's Food Price Update. September/October 2020.

[13] This comprises food, beverages, and tobacco products; wood, paper, and publishing goods; rubber products; and nonmetallic manufacturing commodities (see footnote 7).

[14] Government of Cambodia, MAFF. 2015. *Action Plan for the Implementation of the Cambodia Industrial Development Policy, 2015–2025.* Phnom Penh.

Poverty Reduction

Improved, yet precarious poverty situation. Because of its rapid and sustained economic development, Cambodia has emerged among the global frontrunners in poverty reduction. The country's poverty incidence (i.e., headcount under the national poverty line) significantly declined to 9.5% in 2019 (from 47.8% in 2007).[15] This was accomplished mainly by economic growth driven by increased labor input and employment generation. With its growing working-age population, Cambodia is now enjoying a demographic dividend and has successfully been creating jobs in labor-intensive activities for women and youth.[16] Cambodia's accomplishments in poverty reduction, however, remain highly precarious since many households lifted out of poverty remain close to the poverty line. A study conducted in 2012 revealed that a minor negative shock equivalent to $0.30 per day would trigger a 40% increase in the national poverty rate, which translates to about six million poor people. Nearly 90% of the poor reside in rural areas, and there are sharp regional disparities. The COVID-19 pandemic threatens to undermine Cambodia's progress in poverty reduction. Vulnerable groups such as women and migrant workers are especially at risk. The crisis will lead to an estimated loss of 570,000 jobs and threatens to push an additional 1.1 million people into moderate poverty and 205,000 people into extreme poverty.[17]

Rural–Urban Disparity

Closing, yet significant rural–urban disparity. Rural areas generally lag urban areas in access to basic public goods and services. For example, 2015 data show that access to improved sanitation facilities is still considerably lower in rural areas (30.5%) versus urban areas (88.1%). The national average was 59.2% in 2017. Similarly, in 2015, access to improved water supply was lower in rural areas (69.1%) versus urban areas (100%).[18] Moreover, rural incomes are lower than urban incomes. In 2016, the estimated average monthly household income was KR2.91 million ($717) in Phnom Penh, KR2.46 million ($607) in other urban areas, and KR1.52 million ($374) in rural areas.[19] Rural incomes, therefore, are only about 50%–60% of all urban areas. However, the gap appears to be closing in comparison with 2012 levels of 40%–55%, and this is being accounted for by increases in transfers (i.e., remittances) and by the decline in self-employed agricultural income. In general, Cambodia has seen a steady increase in overseas remittances. Personal remittances increased from 5.0% of GDP in 2010 to 5.6% in 2019.[20] However, because of the COVID-19 pandemic, the country is at risk of losing more than 15% ($420 million) in remittances in 2020.[21]

---

[15] ADB. 2020. *Basic Statistics 2020*. Manila.
[16] The United Nations Population Fund defines a demographic dividend as "the economic growth potential that can result from shifts in a population's age structure, mainly when the share of the working-age population (15 to 64 years old) is larger than the nonworking-age share of the population (14-years-old and younger, and 65-years-old and older)."
[17] ADB. 2020. *Report and Recommendation of the President to the Board of Directors: Proposed Countercyclical Support Facility Loan to the Kingdom of Cambodia for the COVID-19 Active Response and Expenditure Support Program*. Manila.
[18] World Health Organization (WHO) / UNICEF Joint Monitoring Program for Water Supply, Sanitation and Hygiene. The Ministry of Rural Development targets 100% coverage in each area (i.e., rural and urban) by 2025.
[19] Government of Cambodia, Ministry of Planning, National Institute of Statistics. 2017. *Cambodia Socio-Economic Survey 2016*. Phnom Penh.
[20] World Bank. World Development Indicators (accessed November 2020).
[21] A. K. Takenaka, J. Villafuerte, R. Gaspar, and B. Narayanan. 2020. COVID-19 Impact on International Migration, Remittances, and Recipient Households in Developing Asia. *ADB Briefs*. No. 148. Manila: ADB.

Land Availability and Farm Sizes

Decreasing average landholdings and widening inequality in land distribution. Land titles of more than 1 million hectares (ha) of state land were provided to poor people as a measure to formalize existing land entitlements.[22] However, the average landholding size of a rural household is still only 1.3 ha. This size may be sufficient to produce a market surplus for a family of five, but not big enough to propel households into the middle class or provide them with long-term economic security. This trend of decreasing landholding size will likely persist as families continue to divide the farm and parcel out plots to children. Within this overall trend is a polarization of land sizes, with large farms becoming larger and small ones becoming smaller, resulting in increasingly unequal land distribution. Since 2014, agricultural land ownership has not changed much, with 23% of rural households not owning any land and another 15% owning less than 0.5 ha and depending mainly on agricultural wage labor.[23]

This changing distribution pattern of farm sizes has major implications for farm incomes. Extensive crop budgeting and modeling done by the World Bank in 2015 found that returns to land are largely proportional to farm size, with exceptions of maize and dry-season rice production.[24] Hence, the larger the farm size, the larger the returns to family labor and farm income.[25] Returns to land significantly vary across crops. Among different crops, returns to land are significantly higher for vegetable cultivation ($1,394 per ha) and cassava production ($506 per ha) than for the production of other crops.

Labor Availability and Migration

Decreasing share of farm income for the rural population. Between 2003 and 2013, agricultural wage rates rose by a factor of four and converged with rates in other sectors.[26] Since then, many households have diversified from agriculture into more profitable nonfarm activities (often part-time or informal), and the share of wages and transfers (including remittances from growing levels of unskilled labor migration) has risen relative to self-employment in agricultural activities (e.g., cropping, livestock, or fisheries).[27] These nonagricultural sources of income have cushioned the impact of declining agriculture growth post-2012 and are likely to continue to remain important. With Cambodia's relatively small size and good road transport services, it is possible to both operate a farm and undertake some work in urban areas or neighboring countries, especially on a seasonal basis. As of 2018, an estimated 1,000,000 Cambodian migrants were registered in Thailand, and this figure does not include the substantial number of undocumented migrants.[28] These labor outflows in agriculture were expected to continue in the foreseeable future, and then the global pandemic happened. As a result, the economy lost around 400,000 jobs in the first half of 2020, with around 100,000 workers returning to the country. This backflow of mostly rural labor poses an additional challenge to the agriculture sector in terms of how to sustainably absorb the returning migrants.

---

[22] This includes 360,000 ha taken back from Economic Land Concessions as they were improperly awarded to agribusiness firms.

[23] World Bank. 2017. *Cambodia: Sustaining Strong Growth for the Benefit of All.* Systematic Country Diagnostic. Washington, DC.

[24] P. Eliste and S. Zorya. 2015. *Cambodian Agriculture in Transition: Opportunities and Risks.* Washington, DC: World Bank.

[25] Farm sizes have the greatest impact on returns to a day of labor for dry-season rice production, with returns increasing from $6.15 per day for small farms to $8.27 per day for medium-sized farms and soaring to $22.83 per day for large farms. For wet-season rice production, the returns to labor can increase from $2.86 per day for small farms to $12.67 per day for large farms.

[26] This was a period when major primary output prices rose and had a cascading effect on sector incomes. Many farmers with surplus land brought it into cultivation and (to a lesser extent) invested in some improved technology, and higher consumer food prices were offset for many (but not all) poor and landless by higher rural wage rates. It may be noted that the greatest period of Cambodia's national poverty reduction was in the commodity price boom from 2007 to 2012.

[27] By 2014, agricultural income from crop or livestock sales accounted for only 25% of total rural incomes and one-third of the income of the poorest 40% of the population, with the rest coming from wages, household businesses, and remittances.

[28] Government of Cambodia, Ministry of Planning. 2019. *General Population Census of the Kingdom of Cambodia 2019.* Phnom Penh.

Weaving cottage industry. To augment farm income, women in villages engage in cottage goods production such as weaving (photo by Asian Development Bank).

Agriculture—Subsector Assessment, Constraints, and Development Needs

Crop Subsector

Continued dominance of rice production. The country's 4.5 million ha of cultivated land remain largely dominated by rice (70%), followed by subsidiary and industrial crops (20%),[29] rubber plantations (7%), and permanent crops (4%).[30] The cultivated area of rice increased from 3.2 million ha in 2017 to 3.3 million ha in 2018, comprising 2.7 million ha of wet-season rice and 0.6 million ha of dry-season rice. The increased cultivated area of rice resulted in an increase in rice production by around 10.5 million tons in 2017 to 10.9 million tons in 2018 (footnote 7).

Potential rice yield increases with irrigation. Prevailing low wet-season rice yield is the main reason the profitability of Cambodian farming has remained modest; the average yield increase was only 2.6% between 2010 and 2018. Cambodia's wet-season rice yield is about 30% less than that of Thailand's Central Plains region, where varieties other than jasmine and glutinous rice predominate at yields of around 3.8 tons per ha. However, there is scope to close the rice yield gap. Findings show that changing from rainfed to irrigated cultivation in Cambodia could result in annual production increases of up to 40%. For example, the Cambodia Agricultural

---

[29] Industrial crops include sugar cane and tobacco. Subsidiary crops include nonrice cereal and grain crops, root, tuber, and legume crops.
[30] Permanent or perennial crops include banana, cashew, coconut, coffee, durian, oil palm, pepper, mangoes, orange, and other fruits.

Figure 4: Rice Paddy Yield, 2000–2019
(kg/ha)

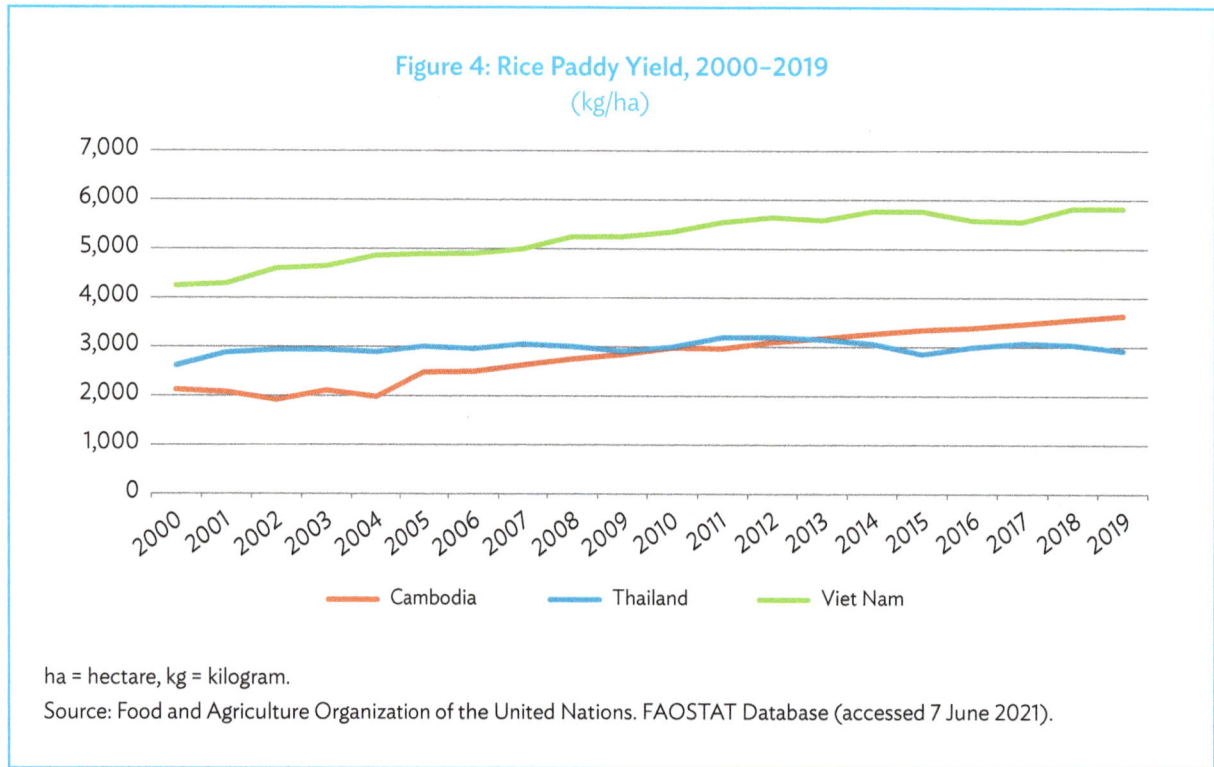

ha = hectare, kg = kilogram.
Source: Food and Agriculture Organization of the United Nations. FAOSTAT Database (accessed 7 June 2021).

Value Chain Program demonstrated on 20 rehabilitated irrigation schemes a total cropping area increase of 42% and an annual output increase from 3.6 tons per ha to 4.8 tons per ha.[31] These yields are comparable to the better arable regions of the Central Plains of Thailand, where farmers achieve 5.7 tons per ha in the dry season, and the Mekong Delta of Viet Nam, where the average yield at a provincial level ranges from 6 to 7 tons per ha (footnote 24). Official yield data from the Food and Agriculture Organization of the United Nations (FAO) show that, on average, Cambodia and Thailand have been achieving similar yields per hectare in recent years, while Viet Nam is still outperforming both countries (Figure 4). As rice farmers are primarily subsistence producers, often with less than 1 ha, they will rely heavily on increased rice yields to generate a marketable surplus.

Increasing rice exports and substantial paddy exports. According to official statistics, total agricultural export volumes are estimated to have risen from 136,853 tons in 2010 to 4,233,532 tons in 2018 (footnote 7). Rice exports increased from 105,259 tons in 2010 to 635,000 tons in 2017 and slightly decreased to 626,255 tons in 2018. These were comprised of aromatic rice (78.8%), long-grain white rice (16.9%), and long-grain steamed rice (4.3%), and were exported to 61 countries by 87 rice-exporting companies. However, most agricultural exports, including rice, remain informal and unrecorded. Although there are no recent data available, informal rice export volumes were estimated at 1.7 million tons of paddy to Viet Nam, and 250,000 tons of paddy and 450,000 of milled rice to Thailand in 2013 (footnote 24).

---

[31]   Cambodia Agricultural Value Chain Program (CAVAC) Phase I (2010–2015), funded by the Government of Australia.

Figure 5: Rice Exports, 2010–2019
($ million, rice milled equivalent)

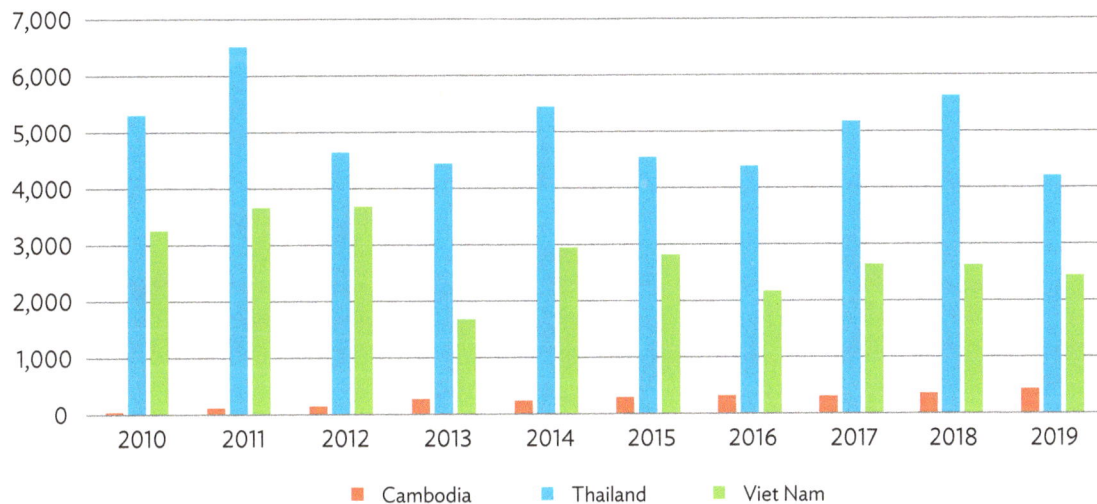

Source: Food and Agriculture Organization of the United Nations. FAOSTAT Database (accessed 7 June 2021).

Official data published by the Ministry of Agriculture, Forestry and Fisheries (MAFF) and FAO disclose the value of rice exports expressed in milled rice equivalent (Figure 5). However, no breakdown between paddy rice and milled rice exports can be found on a yearly basis. Records show that a large portion of paddy rice is being exported as Cambodia has limited facilities to mill their total harvested rice.

Rubber exports with minimum value-added. Despite the major price falls in recent years, the total planted area of rubber has continued to increase from 329,771 ha in 2013 to 436,682 ha in 2018. Rubber production has grown by an annual average of 22.0% (2010–2018) to reach 220,100 tons in 2018; however, production declined by 20.6% to 174,700 tons in 2019. Rubber exports increased from 190,000 tons in 2017 to 217,500 tons in 2018. Rubber is exported in the form of dry sheets or blocks only (i.e., with the absolute minimum amount of processing consistent with producing a saleable commodity).

Increasing subsidiary crop production. Other crops present a mixed picture. Some previously cultivated cash crops (e.g., soybean, peanuts, sesame, sugar cane, jute, tobacco) have seen area and production declines of almost half compared to 2012. However, subsidiary annual crops (e.g., maize, cassava, sweet potatoes, vegetables, and mung bean) saw cultivated area increases from 727,740 ha in 2017 to 1,006,542 ha in 2018. Crop yields of cassava are notably high by regional standards, and maize yields are generally comparable to regional competitors. The total production of these crops was 15,847,801 tons, which increased in volume by about 0.29% between 2017 and 2018.

Increasing perennial crop production. Perennial crop area (including cashew nut, mango, banana, coconut, orange, jackfruit, durian, and black pepper) expanded from 953,597 ha in 2014 to 1,091,000 ha in 2018. Similarly, perennial crop production increased from over 14.7 million tons in 2014 to 16.6 million tons in 2018, up 10.5% over the past 5 years.

Livestock and Poultry Subsector

Significant increase in livestock production. Livestock production has risen by about 50.0% since 2012 and grew by just over 3.0% in the period 2016–2017. Livestock production now constitutes 2.6% of GDP. The total cattle population of 2.9 million is appropriate for 2018 as is the population of buffalo (500,000) in 2018. However, the total pig population has decreased from 2.4 million in 2013 to 1.8 million in 2018, with roughly equal contributions from smallholders and commercial enterprises.

Poor feed and breed quality constraining cattle production. Cattle production is still smallholder-based, with families typically having one to three animals each. The usual route to domestic markets is via small traders or collectors buying at the farm gate (weight and meat composition are estimated by sight, not by actual weighing) for on-selling to local wholesalers, who then arrange trucking to slaughterhouses in provincial towns. Large numbers of live animals are also sold at the borders, especially the border with Viet Nam. Cow productivity is constrained by poor feed quality (usually local grasses). Normally, cattle graze freely in rice fields from February to May and are kept at home and fed rice straw, cut grass, and rice bran mixed with water or banana stems during the planting season. Other production constraints include relatively poor performing or low-weight breeds, poor sanitation, proximity to humans and other livestock species, absence of dietary supplements, low rates of vaccination, and poor village-level biosecurity with uncontrolled grazing practices.

Pig production facing similar constraints as cattle. Families typically have one to five animals. Feed is comprised mainly of kitchen scraps, leftover meat, fish, and bones mixed with rice bran and vegetables. This poor feed regimen means swine generally take 7–9 months from birth to grow to marketable size, and they retain a high fat content at sale. Poor vaccination and disease management practices mean that porcine reproductive respiratory syndrome (or "blue ear"), foot-and-mouth disease, and classical swine fever outbreaks are endemic.

Expanding smallholder and commercial poultry industry. The smallholder poultry population has risen from 21.5 million birds in 2013 to 37.2 million in 2018 (footnote 7). Over the same period, commercial numbers have fluctuated slightly (roughly 7 million birds), but the commercial flock is still double what it was in 2013. In 2016, the MAFF estimated a 21% increase in the number of commercial farms over the previous year, from 2,539 farms to 2,856 during 2014–2015. Poultry production is high in the south and southeast provinces bordering Viet Nam, in the provinces surrounding Tonle Sap Lake, and in the northwest provinces bordering Thailand. Poultry densities directly correspond to human population densities. Industrial chicken production is concentrated in the provinces of Kandal, Kampong Speu, and Siem Reap. Traditional scavenging chicken and duck systems are constrained by disease outbreaks, and transmission risk is greatly magnified through the large live bird markets in urban areas.

Increasing demand for meat but inadequate domestic production. The MAFF's Department of Animal Health and Production estimates meat demand will increase to 300,000 tons annually by 2030. This projection also suggests that only a relatively small proportion (perhaps 20%) of this national demand is presently met through formal domestic supply chains (i.e., from a slaughterhouse onwards). Unsurprisingly, recorded imports of livestock are rising rapidly. Between 2015 and 2016, live poultry imports went up 156%. Growing domestic real incomes and changes in food consumption patterns, as well as demand from neighboring countries and the PRC, will continue to encourage livestock expansion. At present, there are small, improved breeding programs for cattle and pigs. The MAFF data also show there are currently 13 factories producing and selling animal feed made from maize, cassava, broken rice and bran, trash fish, animal bones, and other ingredients. This is replacing feed imports and using domestic raw materials, which were formerly exported to Thailand and Viet Nam.

Animal diseases as threat to livestock industry expansion. One major constraint to future livestock industry expansion will be the increased occurrence of animal diseases, especially in light of the COVID-19 pandemic. Outbreaks of foot-and-mouth disease still occur regularly (there were 100 in 2016), while bird flu has been largely under control in recent years. Recent outbreaks of African swine fever (ASF) are also a major challenge. Cambodia has begun implementing measures on a bilateral and a Greater Mekong Subregion-wide basis to improve sanitary and phytosanitary standards to contain transboundary animal diseases; improving surveillance of veterinary drugs, antibiotics, and feed; maintaining the safety of animal food products; and improving border release procedures.[32]

Fishery Subsector

Fish as an essential source of protein for Cambodians. Fish still provide over 80% of the protein in the national diet, and per capita annual consumption exceeds 63 kilograms (kg) annually.[33] Cambodia's wetlands cover about 30% of the country and support one of the world's biggest, most diverse, and intensive freshwater fisheries. Most harvested fish are from the Tonle Sap and are migratory species. At least 20 species are important for both domestic consumption as well as exports.[34]

Capture fisheries reaching their sustainable limits. Fisheries production has increased by nearly 20% since 2013, but the majority of fisheries output still comes from capture fisheries. Volumes of inland capture fisheries have remained roughly static with 527,795 tons of catch recorded in 2017 and marine capture with just over 121,025 tons. It is widely estimated that national fish catches are only just sustainable, with additional fish catches starting to deplete fish stocks.[35] Various studies on the impacts of Mekong mainstream dams suggest that up to 70% of capture fisheries in Cambodia are under threat and that millions of livelihoods may be affected in the future.[36]

Increasing aquaculture production. Aquaculture production is estimated at 207,443 tons annually in 2017, which is equivalent to 24% of all fisheries production. Production has increased by about 30,000 tons a year in the last 2 years.[37] Inland freshwater aquaculture systems include cage culture, low-input pond systems relying on wild feed supplemented with on-farm products, higher input ponds (typically in peri-urban areas, often on a commercial basis), and rice–fish systems,[38] as well as community fish ponds (e.g., at schools, pagodas). At present, the net incomes from smallholder cage culture are not sufficient to support significant investment, and poor-quality fingerling inputs and the high cost of imported feed are a major hindrance to increased fish farm production.

---

[32] These measures include ADB's support in the form of its trade facilitation project in the GMS. ADB. 2012. *Report and Recommendation of the President to the Board of Directors: Proposed Loans, Grant, and Technical Assistance to the Kingdom of Cambodia and Lao People's Democratic Republic for the Trade Facilitation—Improved Sanitary and Phytosanitary Handling in Greater Mekong Subregion Trade Project*. Manila.

[33] Per capita annual consumption can vary considerably by region and various estimates are shown in FAO, 2011 (footnote 34).

[34] FAO. 2011. *Fishery Value Chain Analysis in Cambodia*. Rome.

[35] A. Brooks, M. Philips, and C. Barlow. 2012. *Fish Supply and Demand in the Lower Mekong with Special Reference to Cambodia*. Final Report for Project FIS/2010/031. Canberra: Australian Centre for International Agricultural Research.

[36] Mekong River Commission. 2011. *Assessment of Basin-Wide Development Scenarios*. Main Report. Basin Development Plan Programme (Phase 2). Phnom Penh and Vientiane; Supreme National Economic Council. 2007. *The Report of Land and Human Development in Cambodia*. Phnom Penh; B. Chan et al. 2020. Temporal Dynamics of Fish Assemblages as a Reflection of Policy Shift from Fishing Concession to Co-Management in One of the World's Largest Tropical Flood Pulse Fisheries. *Water*. 12 (11). 2974; and P. B. Ngor et al. 2018. Evidence of Indiscriminate Fishing Effects in One of the World's Largest Inland Fisheries. *Scientific Reports*. 8. pp. 1–12.

[37] Footnote 8. Besides fish production, aquaculture includes crocodile farming, fish seed production, and hatcheries.

[38] Fish raising in rice fields is increasing but not well documented or reported. S. Freed et al. 2020. Rice Field Fisheries: Wild Aquatic Species Diversity, Food Provision Services and Contribution to Inland Fisheries. *Fisheries Research*. 229. 105615.

Cambodia's inland capture fisheries. The inland capture fisheries in Cambodia are among the largest in the world (photo by Asian Development Bank).

Promotion of community fisheries development. With the cancellation of all previous commercially licensed fishing lots, policy emphasis has now been on community fisheries development.[39] There are presently 475 inland community fisheries and 41 marine community fisheries. This community-based model is suitable for high-value fisher exports but lacks an appropriate value chain structure and organization. Community fishing areas are demarcated with poles, some staffed with rangers, and all have patrol arrangements. There are also 644 fisheries conservation areas nationwide, many congruent with the previously demarcated community fishing areas, and 864 community refuge ponds in 2017.[40] Other conservation activities include planting marine grasses, replanting flooded forests and mangroves, preparing broodstock pools, releasing fingerlings, and conservation of numerous endangered species.

Significant potential for fish exports and value addition. Cambodia exports fresh and processed fish (in smoked and frozen forms) to Hong Kong, China; Malaysia; the Republic of Korea; Singapore; Thailand; and Viet Nam. However, the largest volumes by far are informal or unrecorded flows of fresh or dried fish trucked to Thailand, especially from the Tonle Sap region. There is enormous potential to add value to the national fish catch by processing fish at or near the source. This can range from making sauces, fillets, and packaged or convenience meals. At present, there are few freezing plants in the country. There are no industrial-scale commercial exporters transporting frozen fish products in reefer containers from Sihanoukville or shipping fresh fish from Phnom Penh. Cambodia could tap the competitive European Union and North American markets for high-value species production, but the absence of poor standards makes the markets unattainable for now. As part of the MAFF's Fisheries Administration (FiA) Strategic Planning Framework for Fisheries,

---

[39]   This follows the provisions under Subdecree No. 25 on the Establishment of Community Fisheries, published in 2007.
[40]   Only a few community refuge ponds are active to date.

2016–2030, the FiA will (i) develop a comprehensive range of standards and parameters designed to improve processing, packaging, quality control, and safety; and (ii) implement official registration and inspection procedures to ensure quality assurance and promote trade.[41]

Forestry Subsector

Multiple functions of forests. Forests are important for cash and noncash livelihoods, but their direct contribution to GDP is only 1.4%.[42] With almost 4 million people living within 5 kilometers (km) of forest areas, the share of forest resources in household consumption averages 10%–20%. Households and small and medium-sized enterprises (SMEs) used around 5.5 million tons of fuelwood annually. Additionally, 88% of the population still rely on traditional biomass for cooking, much higher than the regional country average of 58%.[43]

Uncaptured value-addition opportunities in forestry. Official trade data indicate that wood worth $54 million was exported in 2016. Value addition, however, was modest given that the forestry industry in Cambodia is almost negligible, and wood supply has come mainly from economic land concessions, mostly from unauthorized harvested trees in protected areas.[44] The lack of an enabling environment to attract production forestry investments from the private sector or the community is a major constraint.

Lack of formal recognition of non-timber products and forest ecosystem services. The economic and social values of non-timber forest products and forest ecosystem services are not fully reflected in the national accounts, but they provide subsistence for many communities in the form of food, fuelwood, natural medicine, and construction materials for shelter. Forests also support customs, rituals, and traditional practices for many local communities (e.g., Buddhist and animist), which, in some cases, exceed the benefits of commercial timber extraction by more than $200 per ha per year.[45] Furthermore, forests play an important role in supporting a productive fishery sector—e.g., upland forests control erosion, and flooded forests (including mangroves) offer protection for breeding fish. Forests also provide ecosystem services with economic value, such as carbon sequestration, soil conservation, and watershed protection. Studies estimate the average economic value of watershed protection at $321 per ha per year (footnote 45).

Key Sector Constraints and Development Needs

Postharvest Handling and Processing

Limited postharvest handling and processing within the country. An estimated 10% of the total agricultural outputs of Cambodia are processed within the country. This figure has remained relatively consistent since 1998, despite the growth of the economy. Processed agricultural exports made up only 8% of Cambodia's recorded

[41] Government of Cambodia. 2017. *National Strategic Plan for Aquaculture Development in Cambodia, 2016–2030.* Phnom Penh.

[42] Government of Cambodia, Ministry of Planning, National Institute of Statistics. 2019. *National Accounts Data.* Phnom Penh.

[43] Group for the Environment, Renewable Energy and Solidarity (GERES). 2015. *Impact Assessment of Woodfuels Consumption on Deforestation and Forest Degradation in Cambodia.* GERES Southeast Asia Regional Office. Phnom Penh; and GERES. 2016. *Policy Paper: Regulating the Charcoal and Firewood Trade.* GERES Southeast Asia Regional Office. Phnom Penh.

[44] World Integrated Trade Solution. Cambodia Trade Statistics: Products Exports by Cambodia to By Region 2016 (accessed 24 September 2020).

[45] United Nations Programme on Reducing Emissions from Deforestation and Forest Degradation (UN-REDD). 2014. *Valuation of Forest Ecosystems in Cambodia.* UN-REDD Cambodia. Phnom Penh.

exports by value in 2015.[46] The present underdevelopment of the agro-processing sector is related to the main features of Cambodia's industrial structure. This structure consists of (i) limited diversity in the industrial base (i.e., garments, construction, food and beverage processing, and other manufacturing—e.g., bicycles, bicycle and vehicle parts, electronics—which is emerging and export-oriented);[47] (ii) an informal and missing middle-enterprise structure (of the country's enterprises, 97.3% are microenterprises, 2.2% are SMEs, and 0.6% are large enterprises); (iii) relatively immature entrepreneurship (over 42% of enterprises were established between 2008 and 2015); and (iv) an urban-centered industry (68% of large manufacturing enterprises are in Phnom Penh and another 13% in Kandal Province; footnote 46). Production outside the few large enterprises is characterized by low value addition and low levels of technology application. The country lacks a stable or significant cold chain capacity. The overwhelming majority of private ANRRD enterprises still tend to be small, informal, uninformed, and unskilled. They also lack access to finance, accounting and business skills, and information about markets and technologies. Official data on registered SMEs from the Ministry of Industry and Handicraft is inevitably incomplete given the lack of incentives for businesses to register.[48] It does, however, suggest that the majority of formal agricultural and agro-processing SMEs are involved in rice milling and basic processing of rubber and pepper. For convenience foods, only a few enterprises are registered, and they focus on food and drink products, fruit juices, syrups, fish sauce, soy sauce, breads, cakes, and ice making.

Key constraints in agro-processing. Key constraints to increased agro-processing in Cambodia typically include the reliability and cost of electricity, informal payments for permits and documentation, lack of skills and information, difficulties in obtaining development capital, and access to technology and machinery.[49] Many ANRRD private sector activities are unregulated. There is an absence of powerful business associations, with the sole exception of the Cambodia Rice Federation.[50]

Expanding rice milling capacity. Partly because of support from the Cambodia Rice Federation, the country's rice milling capacity has, in recent years, increased remarkably. The milling capacity of larger, modern rice millers increased sevenfold during 2009–2015, from 96 tons per hour (tph) to more than 800 tph.[51] As of 2018, about 200 of Cambodia's over 800 modern rice mills are either medium or large scale.[52] Recent investments in large mills have all been funded by the private sector, of which at least $35 million are joint ventures with foreign investors.[53] Next to rice are (i) cassava processing into flour, starch, or alcohol; (ii) two large sugar cane mills, one opened in 2012 and another in 2016; (iii) paper-based products, including paper and box manufacturing; and (iv) sweets and snacks. There are also 15 feed mills in the country. Drying rubber into sheets or blocks mostly occurs at household and village levels, before being aggregated for export. There are 168 registered slaughterhouses in the country, but only one is of international standard.[54]

---

[46]   Government of Cambodia. 2015. *Cambodia Industrial Development Policy, 2015–2025*. Phnom Penh. Official Cambodian trade data from 2016 indicate that garment exports were worth over $1.1 billion. However, timber, plywood, veneer, and other wood products, rice (husked), rubber, sugar, fishing, and other agricultural products totaled only $50 million (i.e., about 4.5% of all export values). At the same time, recorded imports of drinks (beer and soft drinks), sugar, cooking oil, and other foodstuffs were slightly higher at around $52 million. These figures capture only official trade.

[47]   Harvard University Growth Lab. Atlas of Economic Complexity (accessed 9 February 2020).

[48]   Data on agro-processing SMEs is maintained by the Ministry of Industry and Handicraft, categorized by the size of fixed investment, location, and activity type.

[49]   BDLink (Cambodia) Co., Ltd. 2017. *Agriculture and Agro-Processing Sector in Cambodia*. Phnom Penh.

[50]   The Cambodia Rice Federation is the voice of Cambodia's rice industry. It was established in 2014 by 213 founding members representing rice farmer federations, rice exporter federations, rice exporter companies, rice miller associations, and logistics companies, among others.

[51]   The polishing capacity of mills also greatly improved during 2009–2014, from 72 tph to 520 tph, with a 64% average milling rate.

[52]   A medium mill is defined as having an average milling capacity of 5–10 tph, while a large mill has a capacity of 12–80 tph. There are around 50,000 rice mills overall, including small mills that only operate a few days per year.

[53]   Cambodia's current milling capacity can process all the country's paddy surplus; however, its average milling cost is roughly 30% higher than in Thailand and Viet Nam due to the high costs of electricity and fuel.

[54]   Footnote 8. The slaughterhouse is in Preah Sihanouk.

Farmer with harvested paddy rice. With the increase in Cambodia's rice milling capacity, paddy rice surplus can now be processed (photo by Asian Development Bank Cambodia Resident Mission).

Rising importance of agribusinesses. There is increasing recognition of the potential of agro-processing to contribute to industrial development and national GDP, and this is reflected in the priority given it in the Industrial Development Policy, 2015–2025.[55] A wide range of supporting software measures and capacity building activities have been identified as development needs, which cover (i) improving the investment climate (including laws, regulatory frameworks, and urban and special economic zone planning); (ii) supporting SME development (including improved registration, accounting, and taxation practices; assistance with technology acquisition; and establishment of agro-processing zones); (iii) trade facilitation and export promotion (including implementing the Cross-Border Transportation Agreement); and (iv) industrial standards and property rights protection measures. The Medium Term Plan for the Implementation of Cambodia's Trade Sector-wide Approach (SWAp), 2016–2020 also includes a focus on 10 export products for inclusive value chain development, covering milled rice, cassava, fisheries products, processed food, rubber, and high-value silk products.[56]

Access to Agricultural Inputs and Machinery

Limited availability of improved seed varieties and lack of standards. Most farmers have some access to improved rice seeds but do not apply them to their entire planted area. Survey data (footnote 24) suggest that over two-thirds of households, regardless of geographical area, use local rice seed varieties. Sporadically, through informal networks of middlemen or buyers, farmers are also supplied with seeds of various improved varieties

---

[55] World Integrated Trade Solution. Cambodia Trade Statistics (accessed 9 December 2019). The scale of the potential for import substitution in producing food products from primary outputs may be gauged by the fact that Cambodia imported $732 million worth of food products. About a third of this was from Thailand and only slightly less from Indonesia.

[56] Government of Cambodia, Ministry of Commerce, Department of International Cooperation. 2016. *2016–2020 Medium Term Plan for the Implementation of Cambodia's Trade SWAp*. Phnom Penh.

from Thailand and Viet Nam. The annual supply of quality rice seeds by private companies, cooperatives, and the government increased from less than 9,000 tons in 2015 to 79,200 tons in 2019. However, the supply in 2019 is only 20% of the annual demand of 396,000 tons. Although there are no statistics available, a number of quality rice seeds are known to be imported from neighboring countries. In addition, some quality rice seeds are produced by farmers under the guidance of the MAFF and the provincial departments of agriculture, forestry and fisheries.[57] The Government of Cambodia has not yet ratified the seed policy, including seed quality standards, under the 2008 Seed Law. Moreover, it has yet to make the seed law operational by establishing the essential policies and enforcement mechanisms. For vegetables, the only domestic production of seeds is undertaken by two government research stations, and imports arrive in unrecorded amounts from Thailand and Viet Nam.

Increased fertilizer use. According to FAO statistics, Cambodia's total fertilizer use increased from 38,693 tons in 2005 to 134,053 in 2018. Cambodia imports 100% of its fertilizer. Fertilizer use per hectare of cropland increased from 10 kg in 2005 to 33 kg in 2018.[58] Besides chemical fertilizers, organic fertilizers (such as manure, mulch, and compost) are also used. Cambodia increased imports of fertilizers from about 850,000 tons in 2016 to 1,149,615 tons in 2019.[59]

Increased pesticide use. Since 2015, pesticide use has increased, especially in dry-season rice cultivation and the vegetable sector. More than 3.2 million liters of 100 types of pesticides are used each year.[60] Most of these pesticides come from Thailand and Viet Nam, while smaller quantities are imported from the PRC and the European Union. Informal trade is prevalent, with up to 90% of imported pesticides allegedly either poorly regulated or unregulated. Based on permit applications, the MAFF's latest figures indicate that 81,098 tons of pesticides were imported officially in 2019 (footnote 59). Overuse or misuse of pesticides puts farmers at risk of being poisoned and ecosystems are being polluted.[61]

Increased agricultural mechanization and reduced labor requirement. The type of machinery used depends heavily on agricultural assets, farm size, equipment levels, and availability of services. Tractors are usually used on medium- to large-sized farms, notably in the provinces of Banteay Meanchey, Battambang, and Pailin and typically for land preparation and on upland crops (e.g., maize and cassava). An increase in the use of tractors has been observed in regions where dry-season rice is cultivated, most notably in southern Cambodia. In provinces around the Tonle Sap (Kampong Chhnang, Kampong Thom, Pursat, and Siem Reap), low-lift water pumps are used to irrigate rice fields, often using power tillers to drive the pump's engine. Farmers in provinces close to Phnom Penh (Kandal, Prey Veng, and Takeo) use centrifugal pumps. In general, of the various machinery types, manual tractors are still commonly used in rice and vegetable production, despite the increased uptake of regular tractor use.

---

[57] MAFF General Directorate of Agriculture (Cambodia). 2020. Email message to Takeshi Ueda. 12 November.

[58] FAO. FAOSTAT Database. Fertilizer indicators (accessed 4 November 2020). The composition of nitrogen–phosphate–potash has been changed from 29%–67%–4% to 95%–3%–2%. Recommended composition depends on different factors such as crop type.

[59] Government of Cambodia, MAFF. 2020. *Annual Report for Agricultural Forestry and Fisheries 2019–2020 and Direction 2020–2021.* Phnom Penh.

[60] Cambodian Organic Agriculture Association. 2011. *Pesticidal Plants in Cambodia.* 5th edition. Phnom Penh.

[61] In a survey of over 200 pesticide-using farmers, 88% of respondents had experienced symptoms of poisoning. One reason, apart from the absence of any product labeling in Khmer or any advice via FM radio sources, is the common practice for farmers to mix their own cocktails of chemicals. Besides human illness and even deaths, the improper use of such dangerous chemicals can result in long-term damage to natural ecosystems and their productivity. It is not clear if this situation is improving or worsening.

Figure 6: Usage of Main Agricultural Machinery, 2004–2015
(no. of units)

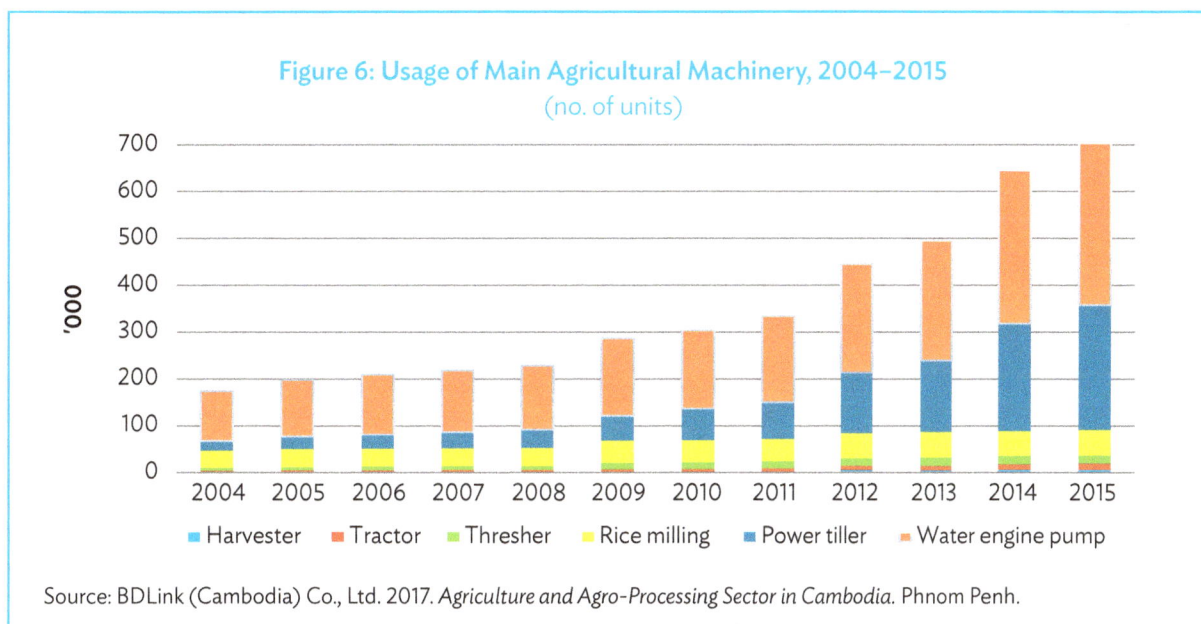

Source: BDLink (Cambodia) Co., Ltd. 2017. *Agriculture and Agro-Processing Sector in Cambodia.* Phnom Penh.

Figure 6 shows the total use of main agricultural machinery in the country. From 2004 to 2015, the use of machines in agriculture has more than quadrupled. The impact of mechanization on labor use is dramatic and indicates what may be in store for agricultural operations in Cambodia. In 2005, 1 ha of wet-season rice production required 73 working days; in 2013, labor use for the same purpose had declined to 48 days. Labor use also declined for maize and dry-season rice as a direct result of increased mechanization (footnote 24).

Figure 7 depicts Cambodia's total employment and value added in the agriculture sector in 2000–2019. A clear downward trend of people employed in agriculture can be seen, while, at the same time, the value added in agriculture has been increasing. This may be another indicator for the increasing mechanization of the agriculture sector in the country.

Figure 8 presents the value-added per worker in the agriculture sector. The performance of Cambodia during 2000–2019 is shown alongside trends in neighboring countries.

Access to Markets

Dominant conventional sales of paddy at farm gate. The majority of farmers continue to sell their main crops (i.e., paddy)[62] to traders or collectors in their village just after harvest.[63] At this time of relative abundance, farmers tend to be price takers and their returns are therefore depressed—although, the system as a whole is reasonably competitive, and spatial and temporal price differences are minimized. For cash sales of rice, these traders or collectors are small, highly mobile private enterprises, and sales average about 1.5 tons per year. One constraint

---

[62]   Rice for domestic consumption is mostly stored at home as paddy, taken to a mill to be processed when needed, then returned to the farm or household as edible rice. The milling fee is typically paid in the form of a proportion of milled rice or bran retained by the miller.

[63]   In Cambodia, there is evidence of farmers making direct deliveries to markets. However, in all cases, the most common logistic activity is the collection of produce by a trader or middleman at farm level, and then onward delivery to markets and final customers in small, medium, or large trucks. ADB. 2016. *Wholesale Fresh Food Markets in Selected Countries in Southeast Asia.* Reconnaissance study. Manila (TA 8163-REG).

Figure 7: Employment and Value Added in Agriculture, 2000–2019

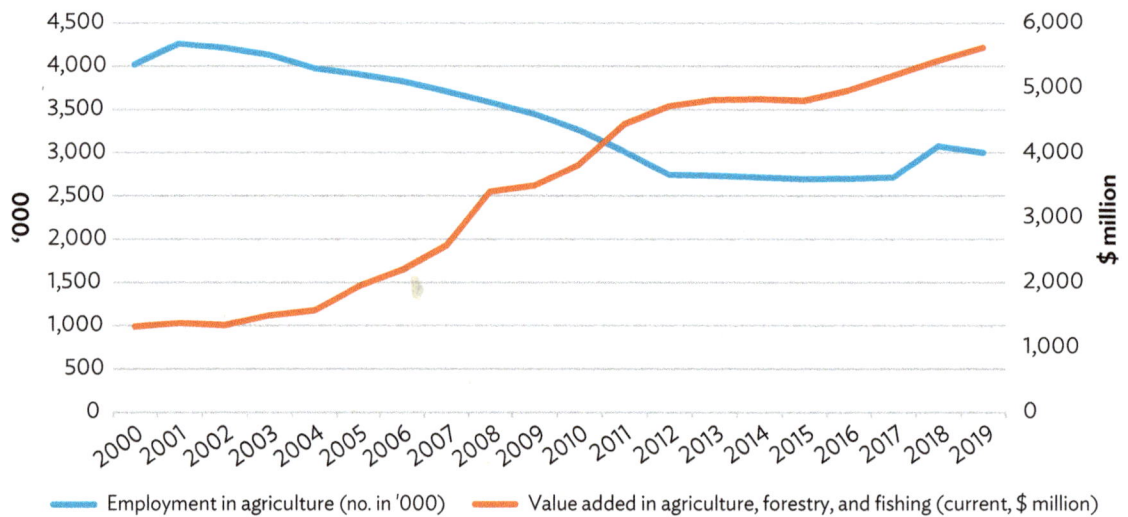

Sources: Asian Development Bank. *Key Indicators for Asia and the Pacific*. Manila (2000–2020); International Labour Organization. ILOSTAT Database (accessed 10 November 2020); and World Bank. DataBank (accessed 10 November 2020).

Figure 8: Value Added per Worker in Agriculture, Forestry, and Fishing, 2000–2019
(constant 2010, $)

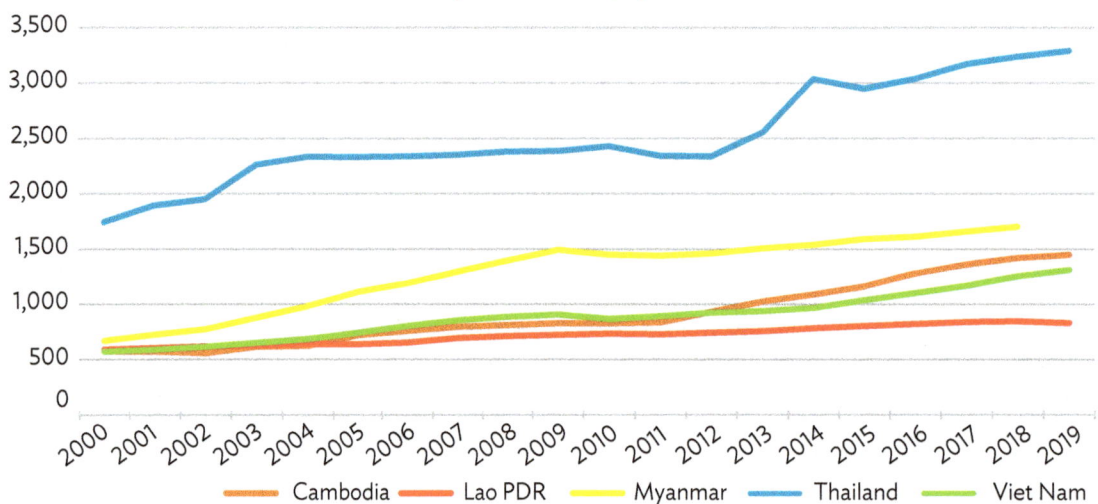

Lao PDR = Lao People's Democratic Republic.
Note: The most recent available data for Myanmar are only up to 2018.
Source: World Bank. DataBank (accessed 16 November 2020).

to increasing value addition is that varieties are mixed and only relatively small volumes of pure varieties can be found. Collectors will typically combine purchases and sell them on to traders who may be specialist rice traders or involved with other commodities and goods.

Institutional purchases of rice for the army, the police, and the national food reserve are also significant. Both traders and wholesalers are responsible for arranging large-scale informal exports of rice across Cambodia's borders into Thailand and Viet Nam. Retailers of rice in Cambodia comprise rice shops (including those which specialize in wholesale procurement) and market stalls. Although farmers may receive about 70% of the retail price for their crop, their returns are considered small after deducting input costs and labor.

Market access for vegetables. The situation with vegetables and fish marketing is similar. The largest share of products (around 90%) are typically sold to traveling collectors or traders or direct to wholesalers.[64] Collectors will mainly sell the output at local district or provincial wet markets or sell to retailers. Some farmers may also sell at local retail markets or act as street vendors. For crop sales that have gone to wholesalers, these may be combined with crops from other areas of Cambodia or with imports and then sold to grocery stores and supermarkets.

Increasing importance of access to international markets. The importance of access to and the conditions within international markets of the ANRRD outputs is critical.[65] Current forecasts of world demand and prices for Cambodia's present export mix suggest a commodity boom will not be repeated. Since August 2020, Cambodia's access to European Union markets has been affected by partial withdrawal of preferential treatment under the "Everything but Arms" program.[66] According to a World Food Programme market update, the COVID-19 pandemic caused some initial disruptions to food supplies due to border closures in early 2020. These initial fluctuations have since then been reversed, and agricultural markets were reportedly functioning well in the second half of 2020.[67]

Transport and Logistics

Weak and deteriorating rural road network. Roads are vital to both economic and social development, particularly in districts where there are no other available means of transport. The rural road network, on which most of the ANRRD primary production depends for inputs supply and market access, continues to deteriorate and constrains sector efficiency and growth.[68] Cambodia's transport infrastructure quality index decreased from 3.1 (2010) to 2.1 (2018) out of a total score of 7.0, and the country scores particularly low on infrastructure.[69] Among the factors likely to explain this fall in the quality index are poor standards and weak institutional capacity for road maintenance and lack of financing for maintenance and management resulting in problems

---

64   ADB. 2005. *Enhancing Agricultural Project Economic and Financial Analysis: Cambodia, Lao People's Democratic Republic, and Viet Nam.* Manila.

65   Increases in the ANRRD sector incomes up to 2013 were due, in large part, to a world commodity price boom. In Cambodia's case, the boom was amplified by the operations of the rice-pledging scheme in Thailand, which absorbed large amounts of paddy, mainly from the northwestern provinces.

66   *European Commission. 2020. Cambodia Loses Duty-Free Access to the EU Market Over Human Rights Concerns. 12 August.* The Everything but Arms initiative is a trade incentive scheme under the European Union's Generalized Scheme of Preferences that grants duty-free and quota-free access of imports from least developed countries, except arms and ammunition.

67   World Food Programme. 2020. *Cambodia Food Price Update–August 2020.* Rome.

68   The primary road network has (i) 2,254 km of national paved roads that connect the country with its borders, the GMS, and the Association of Southeast Asian Nations network; (ii) 5,007 km of inland national roads, of which 70.4% are paved; and (iii) 9,031 km of provincial roads, of which only 30.4% are paved. The Ministry of Public Works and Transport manages the national and provincial roads, while the Ministry for Rural Development is responsible for the remaining network of about 45,241 km of rural roads (roughly 74.0% of the total road network). Until 2017, rural roads were still principally laterite or earth roads; those paved with either a single-bituminous surface treatment, a double-bituminous surface treatment, or reinforced concrete comprised just 4.8% of rural roads.

69   World Bank. Logistics Performance Index (accessed 30 September 2020). The index is a composite of customs, infrastructure, international shipments, logistic competence, tracking and tracing, and timeliness.

Importance of rural roads. In many parts of Cambodia, rural roads are the main access for agricultural inputs supply and for transporting products to markets (photo by Asian Development Bank).

of overloading, poor safety systems, and rapidly growing traffic volumes.[70] In addition, the inadequate capacity of private contractors, flaws in design, and weaknesses in construction methods contribute to the situation. The effects of climate change are now being seen in the increased incidence of road flooding. Largely because of these constraints to the performance of the national road network, Cambodia's domestic transport costs remain higher than its regional competitors. Road improvements, more trucks, and promoting competition among local transport companies helped bring down Cambodia's transportation costs from an estimated $15 per 100 km per ton in 2009 to around $10–$13 per 100 km per ton in 2015. However, this is still higher than in Viet Nam ($7 per 100 km per ton) and Thailand ($5 per 100 km per ton).

Increasing fund allocation for rural road maintenance. Given the need to address the systemic issues with road maintenance and as part of the ongoing public financial reform program, the government aims to increase maintenance funding by 5.5% per year to improve the sustainability of the road network. Although a road fund is an appropriate mechanism to ensure annual funding, the government prefers to allocate funding from central government resources.[71] The Road Asset Management Project, cofinanced by ADB, the World Bank, and the Government of Australia, contributes to improving the maintenance of the national road network.[72] Similarly, the Rural Roads Improvement Project (RRIP II), financed by ADB, is systematically improving the maintenance

---

[70] There are no feasibility study requirements for project selection nor minimum standards for project implementation set by most line ministries. Because of this lack of internal capacity, large infrastructure projects such as roads and canals are often delegated to development partners or donors. This clearly poses challenges for the provision of public infrastructure of all types (e.g., roads, irrigation, industrial parks, special economic zones) to ensure economic competitiveness. World Bank. 2017. *Cambodia Economic Update—Investing in Public Infrastructure and Services*. Phnom Penh.

[71] A road fund would involve an allocation of resources from specific sources (e.g., vehicle or fuel taxes).

[72] ADB. 2007. *Report and Recommendation of the President to the Board of Directors: Proposed Loan and Administration of Grant from the Government of Australia to the Kingdom of Cambodia for the Road Asset Management Project*. Manila (Loan 2406-CAM).

Night Market in Phnom Penh. A wide array of Cambodian goods, including food and agricultural produce, is sold in night markets (photo by Asian Development Bank).

of the rural road network.[73] Ways to involve the private sector in rural infrastructure financing are now being investigated by the government.[74]

Declining inland water transport. Of the 1,750 km of inland waterways in Cambodia, all-year navigation is limited to 580 km on the Mekong River and its tributaries. The primary use is for containers, bulk general cargo, and petroleum. Although barges operate along the river to Phnom Penh, inland waterway transport has been declining in recent years as cargoes have switched to the more flexible and cheaper option of roads. By comparison, both Thailand and Viet Nam heavily rely on their waterways for transporting rice and other agricultural produce. In Viet Nam, water transport costs $3 per ton, while in Cambodia, it would cost $15–$17 per ton to move the same distance by truck (footnote 23).

Improving logistics performance. The government has, in recent years, embarked on significant improvements in trade facilitation and logistics, but Cambodia's export charges remain considerably higher compared with other countries. In terms of its ranking on the World Bank's Logistics Performance Index (LPI), Cambodia has consistently improved, rising to 89th place in the aggregated LPI (for 2012, 2014, 2016, and 2018) after ranking 129th in 2010, with its score increasing from 2.36 to 2.66.[75] The country also ranks fourth among the top 10 performing low-income economies. This is mainly due to the series of reforms the government introduced, including (i) the ratification of a customs law, (ii) the approval of a risk management strategy at customs, (iii) the implementation of the Automated System for Customs Data following international standards, (iv) the adoption

---

[73]   ADB. 2014. *Report and Recommendation of the President to the Board of Directors: Proposed Loan and Administration of Grants to the Kingdom of Cambodia for the Rural Roads Improvement Project II*. Manila (Loan 3151-CAM).

[74]   ADB. 2018. *Report and Recommendation of the President to the Board of Directors: Proposed Loan and Grant to the Kingdom of Cambodia for the Rural Roads Improvement Project III*. Manila (Loan 3678-CAM).

[75]   World Bank. Aggregated LPI 2012–2018 (accessed 18 September 2020).

of a single-stop inspection system at the border, and (v) the simplification of transit operations with neighboring countries through the Cross-Border Transportation Agreement.[76] However, according to a World Bank report, informal fees increase the total transport and logistics costs (footnote 23). For example, there are numerous reports of unofficial payments being routinely collected from trucking firms at checkpoints and weighbridges. Accounts from interviews with private sector stakeholders reveal that forwarding charges in excess of the official customs processing fee ($15) can be up to $200 per container, which is equivalent to the difference in the average cost of exporting a container between Cambodia (about $800) and Thailand or Viet Nam (about $600).

Market infrastructure in need of improvement. Cambodia's physical market infrastructure is in a poor state and requires significant improvements. A recent study of regional wholesale fresh food markets concluded that Phnom Penh's day markets for fruit, poultry, and meat products are outdated, crowded, unhygienic, and need upgrading or relocation.[77] Meat cuts are prepared from carcasses then bagged and moved by hand or on motorbikes. Meat stalls are situated close to fruit and vegetable sellers, with obvious risks of contamination, and waste and rubbish are collected and moved along narrow alleyways that include meat stallholders. The restrictions on opening hours for night wholesale markets mean that stallholders must set up on pavement sites, thereby causing noise, waste, green litter, and traffic disruption for residents. Large trucks unload quantities of fresh produce on side streets, and the streets are blocked with parked vans and trucks unloading and reloading each night. At present, there is little transparency on rents and market finances, and there are no signs of any private or public investments to upgrade existing markets or to relocate or invest in new markets. The situation in other large urban centers (e.g., Battambang, Siem Reap) is similar in qualitative terms, if slightly different in scale to Phnom Penh.

Industrial zones as drivers for industrial development. Recognizing the need to improve the overall logistics performance for the country and to better integrate with regional and global agro-based value chains, the Industrial Development Policy, 2015–2025 highlights the following: (i) the establishment of industrial zones to facilitate economic linkages, critical mass, and competitiveness; (ii) the development of specialized economic corridors (e.g., the Kampong Cham Province–North–West industrial corridor for agricultural processing zones); and (iii) the construction of new industrial parks and industrial clusters, with at least some of them oriented toward agro-processing.[78] Meanwhile, in line with these government policies, numerous private investments are at various stages of formulation or implementation, some at a significant scale.[79]

Access to Information and Communication Technology

The government has undertaken several policy-level interventions toward digital development in recent years. It developed its Information and Communication Technology (ICT) Masterplan 2020[80] and drafted the Cambodia e-Government Master Plan, 2017–2022.[81] This has helped the country achieve notable progress in

[76]   Government of Cambodia, Ministry of Public Works and Transport. 2020. *Interim Master Plan on Intermodal Transport and Logistics Connectivity.* Phnom Penh.

[77]   ADB. 2016. *Wholesale Fresh Food Markets in Selected Countries in Southeast Asia.* Reconnaissance study. Manila (TA 8163-REG).

[78]   Government of Cambodia. 2015. *Cambodia Industrial Development Policy, 2015–2025.* Phnom Penh.

[79]   For example, in late 2016, the Chinese Tianrui group signed a memorandum of understanding with the MAFF to build a $2.1 billion special economic zone on a 300-hectare site in Kampong Speu Province. The orientation is mainly to serve the group's 25 Chinese supermarkets, but the domestic market would also be served by this investment. A large part of the initial funding was to go to contracts for farmers, who would seed, grow, nurture, and harvest a variety of crops in the area. In addition to storage facilities, a phytosanitation plant was to be established to study and observe the most common plant diseases in Cambodia. O. McCall. 2016. Chinese-Funded Special Economic Zone to Rise in Cambodia. *Born2Invest.* 24 October.

[80]   Government of Cambodia. 2014. *Cambodian ICT Masterplan 2020.* Phnom Penh.

[81]   This master plan has not yet been adopted.

digital infrastructure and service delivery. However, the country still ranks at the lower end in digital adoption and technological readiness, particularly because of low digital adoption by businesses and the government.[82]

Importance of information and communication technology in agriculture value chains. ICT can play an important role in linking stakeholders along agriculture value chains including farmers, agribusiness, consumers, and government agencies. ICT can make available aggregated data on agriculture production and demand, among other information, to all value chain actors, and facilitate access to market information, extension advice, and communication between farmers and agribusiness entrepreneurs. The digitization of information will facilitate the formation of value chain platforms through which value chain stakeholders can exchange information, services, and products.

Multiple ICT interventions are currently planned by various ministries that could have an impact on efficiencies along agricultural value chains. The Techo Startup Center is a government-funded organization that seeks to implement an open digital platform for agriculture value chain development through the Khmer Agriculture Suite (KAS digital platform). This platform is currently under development and will provide training in digital literacy and the use of existing digital applications for value chain actors.

Access to Finance

Increased access to finance in rural areas. Cambodia has witnessed one of the fastest capital-deepening periods in Southeast Asia. Domestic credit to the private sector as a percentage of GDP skyrocketed from 27.6% in 2010 to 114.2% in 2019.[83] The expansion of microfinance institutions (MFIs) has been remarkable, with both assets and credits of the MFIs growing at rates of more than 40%–50% annually since the global financial crisis.[84] In contrast to the period prior to 2010, credit sources for rural areas are now widely available. Commercial banks, MFIs, and community savings now reach over 90% of villages. ACLEDA Bank has grown to become the largest MFI in Cambodia and now has commercial bank status. Community savings and loans institutions are also being established in more villages. The sources of credit used vary by farm size and type of operation. Commercial growers of rice, maize, and cassava tend to use both MFIs and commercial banks, while smaller farmers typically prefer community savings and loans institutions as well as traditional moneylenders. Among the smaller rice and vegetable growers, less than 10% use commercial services. Despite the increasing availability of formal credit sources, the proportion of small farmers reportedly taking loans from moneylenders appears to be increasing.

Access to finance as major bottleneck for farmers and agribusiness. Despite the increased access to finance in rural areas, credit to agriculture and agribusinesses together accounted for only 9.4% of formal finance in Cambodia in 2018. Access to finance remains the major bottleneck for Cambodian farmers and agribusinesses. First, they cannot meet the high collateral requirements. Despite the government's promotion of cash-flow-based lending, financial institutions require collateral because of poor cash flows in agribusinesses. Second, financial institutions tend to charge high interest and provide short-term loans because of the risks associated with agriculture and agribusinesses that are small and informal and deal with price-volatile agricultural products. The lack of credit history and low financial literacy are other reasons for their poor financial access. There is some indication that, because of the COVID-19 pandemic, access to credit and finance has been restricted. Addressing constrained access to finance is critical to spurring post-pandemic recovery.

---

[82] World Bank. 2018. *Benefiting from the Digital Economy: Cambodia Policy Note.* Washington, DC.

[83] National Bank of Cambodia.

[84] Compared with only $426 million in outstanding loans and 0.9 million borrowers by the end of 2010, approximately 50 MFIs and 8 micro-deposit-taking institutions lent a record high of $2.9 billion to 2 million borrowers in 2015.

Extension Services

Uneven coverage of extension services. Agricultural extension is provided by the government, the private sector, and nongovernment organizations (NGOs). Survey evidence suggests that about 70% of all villages receive some type of extension inputs, but much of the advice (largely on rice) remains generic and is not tailored to local circumstances (footnote 24). The various extension providers adopt different methods. The public sector emphasizes pluralistic advisory services and applies approaches such as farmer field schools, demonstration plots, mission-based services, farmer field days, study tours, mass media campaigns on television and radio, and awareness-raising materials (e.g., posters, booklets, and leaflets). Individual, donor-supported projects generally supplement these approaches.[85] NGOs tend to use issue-based farmer group training, input trials, and multistakeholder platforms for delivering extension services to farmer groups. These typically involve all actors, from production to market, in a discussion of constraints, challenges, and opportunities for production improvements and business arrangements. Increasingly, extension services are delivered by the private sector, which commonly uses field trials and focuses on the provision of inputs directly to farmers. Specific approaches depend on the business arrangements between farmers and companies or their agents and may be in the context of some type of contract farming.

Shift to demand-driven, localized public extension services. Public extension services are in the process of being reoriented, with a shift in strategy from the historical supply-driven extension-for-production approach to a demand-driven extension-for-market approach, otherwise known as the competition-and-sustainability approach. Consistent with public reform programs, (i) extension activities are currently being developed to better address the agriculture development needs and priorities at national, provincial, and district levels; and (ii) services are being decentralized by strengthening staff capacity at district and commune levels, with particular focus on planning and value chain facilitation.[86] By 2020, the district agriculture office, together with the district technical support team, will be primarily responsible for providing technical support and extension training at the local level. Commune extension centers are also being developed at nearby locations to bring extension services to farmers and their organizations. Nevertheless, with the limited number of public extension workers (one public extension worker per 5,000 farmers), further streamlining of service delivery and complementing government extension services with private sector services may be options to address the needs of farmers who are trying to meet consumer demand for more diversified and higher quality agricultural products. Either in the public or private provision of extension services, ICT can play an important role in service delivery.

---

[85] Within the Agriculture Services Programme for Innovation, Resilience and Extension (ASPIRE), the MAFF's Department of Agriculture Extension has established an extension hub for improving access to extension and advisory services. The MAFF also established farmer associations (including the Improved Group Revolving Fund), agricultural cooperatives, agribusiness clusters (through the Project for Agriculture Development and Economic Empowerment), research stations, and agricultural development centers to serve agricultural research and innovative technology transfer.

[86] Specific objectives and targets are being formulated in each district in collaboration with agricultural researchers, rural development officers, market researchers, and climate change specialists, as well as with farmer organizations, community-based organizations, NGOs, banks, and the private sector.

Women vegetable farmers in Cambodia. Women farmers are working within farmer cooperatives to build their capacity to grow and sell more agricultural products, and thereby earn more (photo by Asian Development Bank).

Farmer Cooperatives

Growing importance of farmer cooperatives. Strengthened farmer organizations are essential to improve production and marketing arrangements. Cooperatives development is a key part of the MAFF's strategy and is taken as a prerequisite for contract farming.[87] The number as well as importance of agricultural cooperatives is growing.[88] As of 2019, the MAFF recorded around 1,200 registered cooperatives, with just under 122,000 members (some 60% of members being women). In line with the government strategy, the development of cooperatives remains a focus of support for several development partners, including FAO and the Japan International Cooperation Agency, in the form of stand-alone projects, and for other development partners, such as the International Fund for Agricultural Development (IFAD) and the World Bank, within wider value chain investments. Numerous projects supported by development partners advocate farmer self-help groups. One example is ADB's Emergency Food Assistance Project, where 42 out of over 400 self-help groups were transformed into registered cooperatives undertaking both input supply and collective produce marketing as well as providing seasonal credit.[89]

---

[87] The MAFF's vision is the empowerment of producers working together to develop resource sustainably, learn lessons, and share knowledge. The MAFF works closely with larger commercial operators and encourages them to be fully inclusive, cooperative, and competitive in an open market for the attainment of producer power in their partnerships with traders through formal contract agreements. These organizations will be supported by public extension services and appropriate private sector assistance. Government of Cambodia, MAFF. 2015. *Agricultural Sector Strategic Development Plan, 2014–2018*. Phnom Penh.

[88] The Royal Decree on Cooperatives in Cambodia was issued in 2001, while the law concerning agricultural cooperatives was passed in 2013.

[89] ADB. 2008. *Report and Recommendation of the President to the Board of Directors: Proposed Loan and Asian Development Fund Grant to the Kingdom of Cambodia for the Emergency Food Assistance Project*. Manila.

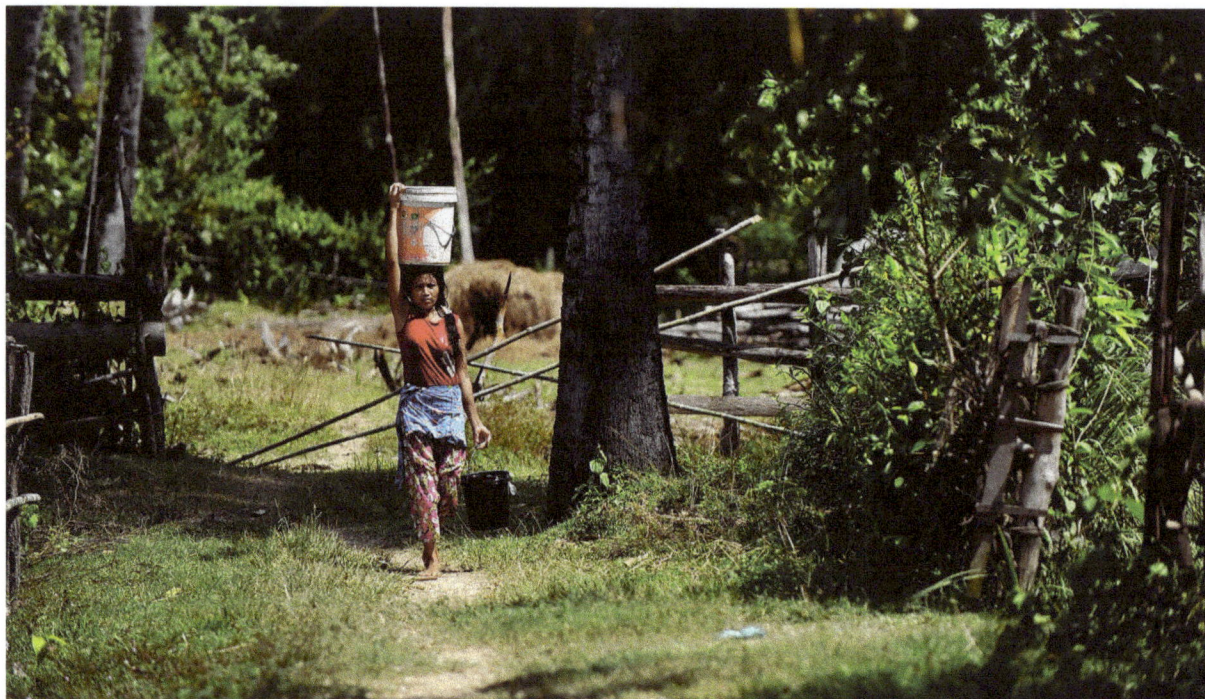

Woman carrying drinking water. In the rural areas of Cambodia, women are the ones primarily tasked with fetching water (photo by Asian Development Bank).

Natural Resources—Subsector Assessment, Constraints, and Development Needs

Water Resources

Seasonal water scarcity. Approximately 472,000 million cubic meters (mcm) of water—and 105,000 mcm of dry-season water—flow through the country annually. Some 30% of this comes from rainfall. National water use is only about 10,000 mcm annually and 7,000 mcm during the dry season for all purposes including irrigation, industry, and domestic use. This is equivalent to an estimated 2%–7% of available water.[90] Water scarcity results primarily from a lack of water storage capacity and limited access to water during the dry season and parts of the wet season, and not from the available volume of water.[91]

Limited potential for water harvesting and storage facilities to improve water availability. Cambodia's flat topography, especially in the Tonle Sap river basin group, where 35% of the irrigation schemes are located, means there are limited areas for large-scale water storage facilities. Most water is presently stored using water harvesting embankments across low-lying areas. Water control is generally difficult, and losses from evaporation,

---

90   ADB. 2014. *Cambodian Water Resources Profile 2014*. Consultant's report (Egis-eau). Manila (Loan 2673-CAM and TA 7610-CAM).
91   Cambodia has 39 river basins that combine to form the five river basin groups. The most severe water shortages occur in (i) the 16 river basins that drain into the Tonle Sap Lake; and (ii) the 3S river basins (i.e., Sekong, Srepok, and Sesan), of which the upper catchments are in Lao People's Democratic Republic and Viet Nam. For the upper Srepok and Sesan river basins located in Viet Nam, the natural flow regime has been highly modified through hydropower development; hence, dry-season water in Cambodia is dependent on water releases from reservoirs in Viet Nam. For the Tonle Sap river basin group, all the river basins are located within Cambodia—except for Stung Mongkol and Stung Sisophon river basins, where the upper 20%–30% of their catchments are in Thailand.

seepage, and poor management are high. Extreme flooding, flash floods, and droughts are regular events, causing high rates of death, extensive crop and water infrastructure damages, increased spread of waterborne diseases, disruption to social and economic activities, and displaced populations. The effects on underlying agricultural production conditions are significant.[92] Groundwater is available across most of the country, primarily for drinking and subsistence farming. Reports are increasing of farmers installing tube wells to support small-scale irrigation and to provide a supplementary water source for dry-season cropping. Increasing national water use productivity requires better water harvesting and storage capacity on farms as well as more efficient water use (e.g., through the use of drip irrigation systems for vegetable and fruit cultivation).

Good surface water quality. Based on the findings from the Mekong River Commission monitoring program, overall salinity levels of surface water are generally low and not yet restricting agriculture, and concentrations of pollutants have only moderate to low impact on aquatic ecosystem health.[93] However, of some concern are the increasing pressures from human activities on water quality on the western side of the Tonle Sap and in the Mekong Delta region, which are attributed to high population densities and increasing pesticide use for farming.[94]

Significant investment needed for the rehabilitation and construction of irrigation schemes. Cambodia has approximately 2,480 irrigation schemes, but the majority are not fully operational, thereby constraining agricultural productivity.[95] Water delivery within many of these schemes is often inefficient and does not reach all farms, especially plots furthest from the primary canal network. Out of 4.5 million ha of arable land, including 3.2 million ha of rice-cultivated area, irrigation covers around 1 million ha (22%). The national annual irrigated area is around 1.5 million ha (i.e., 1 million ha wet-season cropping and 0.5 million ha dry-season cropping). Over $1 billion has been invested in the irrigation sector since 2010. Most of this investment has been to support the rehabilitation of existing irrigation schemes and the construction of a small number of new and large irrigation schemes and reservoirs, but not on ensuring the full completion of irrigation schemes. The quality of construction also varies considerably as there are no formal quality assurance procedures for the selection, design, and construction of schemes.

Neglected operation and maintenance requirements undermining productivity and sustainability of irrigation schemes. Another major constraint to irrigation productivity is that investment has not been sufficiently cognizant of operation and maintenance (O&M) requirements to support the long-term sustainability of irrigation schemes. Studies show that rice yields can increase by 125% per annum by shifting from rainfed to irrigated agriculture, but this is dependent on having year-round availability of water. Farmer water user communities (FWUCs) on irrigation schemes have been promoted by various government initiatives and development partner investments (including ADB irrigation projects) in recent years. But their performance has been variable, and the legislative framework governing their operation and organization remains incomplete. In many cases, FWUCs have failed to sustain irrigation service fee collection and to fully interact with the Provincial Department of Water Resources and Meteorology (PDWRAM) to maintain physical infrastructure and

---

[92] The most fertile soil regions are the alluvial floodplains around the Tonle Sap Lake and the upland regions to the west of the Lake and in the upper Mekong Delta region. All these areas are prone to flooding. Almost half of Cambodia's land area is sandy soils (Acrisols) that have low nutrients and a high permeability level and are prone to hard-setting, making cultivation difficult. Nutrients are susceptible to being washed out during the wet season, causing farmers to apply high amounts of fertilizer during the dry season.

[93] Mekong River Commission (MRC). 2008. *Bio-monitoring of the Lower Mekong and Selected Tributaries, 2004–2007*. MRC Technical Paper No. 20. Vientiane; MRC. 2016. *2014 Lower Mekong Regional Water Quality Monitoring Report*. MRC Technical Paper No. 60. Vientiane; and MRC. 2018. *2016 Lower Mekong Regional Water Quality Monitoring Report*. Vientiane.

[94] MRC. 2018. *2016 Lower Mekong Regional Water Quality Monitoring Report*. Vientiane. From 2000 to 2016, total phosphorus and chemical oxygen demand levels increased slightly, while nitrate-nitrite, ammonium, and dissolved oxygen levels remained relatively constant.

[95] Approximately 1,730 irrigation schemes have the potential to be rehabilitated, 250 schemes are considered almost fully functional, and 500 schemes will likely need to be written off at some point. Figures are from the Government of Cambodia, Ministry of Water Resources and Meteorology (MOWRAM). Cambodia Irrigation Scheme Information System.

Irrigated agriculture in Cambodia. The Asian Development Bank is helping to improve and rehabilitate the irrigation systems in Cambodia (photo by Asian Development Bank Cambodia Resident Mission).

water supply. This is the main reason why few irrigation schemes are presently achieving double-cropping on a sustained basis.[96] The capacities of FWUCs and the PDWRAM are rarely assessed in detail at the design stage, and suitable training programs are likewise only rarely established to build the skills needed to undertake O&M responsibilities.[97]

Limited climate and surface water resources information. The nationwide hydro-meteorological monitoring network is limited, with poor water data records that are often incomplete and less than 20 years old. There are 123 hydrological monitoring stations across the country, but only about 40–50 of them are operational. Most stations provide only water level observations with some discharge measures in the Mekong mainstream.[98] Initial water accounting work and surface water resources assessments using remote sensing data and hydrological models at the river basin scale have been supported by ADB. Hydro-meteorological and water resources monitoring needs to be further strengthened to enhance climate data and water resources information that are essential for developing water resources management plans and equitable water allocation rules at river basin levels, forecasting and early warning for floods and droughts, and informing investment decisions.

---

[96] Notwithstanding these problems, Cambodia does have regions of water productivity that are above global averages. The 3S river basin group performed the best, achieving up to 1.35 kg per meter water productivity for irrigated rice (only slightly lower than Viet Nam's rate at 1.36 kg per meter). The global average is around 1.10 kg per meter. However, annual rice production estimates for double-cropping in Cambodia are up to 7 tons per ha, whereas the global average is around 12 tons per ha (6 tons per crop).

[97] As a guide to future interventions, the Australian government-funded Cambodia Agricultural Value Chain Program (CAVAC) is one of a few irrigation investment programs that has demonstrated how to establish sustainable, productive, and profitable irrigation schemes in Cambodia. Rehabilitated schemes are now achieving increased annual production value allowing full pay-back of all costs within 2–3 years. Most CAVAC FWUCs achieve a 95% irrigation service fee collection rate, covering routine (annual) and periodic (5–10 years) O&M, making them independent of the public recurrent budget.

[98] Government of Cambodia, MOWRAM. 2019. *National Water Resources Management and Sustainable Irrigation Road Map and Investment Program, 2019-2033.* Phnom Penh.

Need for groundwater resources monitoring. Groundwater information is typically from a few localized research projects. This is a concern considering the increasing reports of groundwater pumping to supplement dry-season water shortage across most regions of the country. A water accounting study by ADB and the IHE Delft Institute for Water Education provides the most recent information on groundwater changes.[99] This study found that groundwater stress varies spatially (i.e., between river basin groups) and temporally (i.e., between 2004, 2007, and 2008). In 2004, the driest year, groundwater withdrawal was almost double the recharge in four of the five river basin groups. This pattern may be related to rainfall–runoff rates, with higher runoff rates and lower infiltration rates during the wet season.

Donor shift to support modernized, whole-of-system water resources management approaches. Overall, a consensus among development partners is that the need for water resources has been shifting from investment focusing only on the construction of irrigation infrastructure to investment supporting a modernized, whole-of-system approach to water resources management.[100] The whole-of-system approach is based on the assumption that infrastructure should be developed with the aim of increased agricultural productivity and profitability over the long term and should, therefore, respond to agriculture needs and strategies such as crop diversification. Greater emphasis also needs to be placed on quality assurance at all stages of establishing fully functional irrigation schemes, including (i) having a secure water supply to support dry-season cropping, (ii) ensuring the quality of design and construction of irrigation schemes, and (iii) supporting well-functioning FWUCs that are not solely dependent on O&M subsidies from the government. These measures, in turn, need to be supported by improved decision-making, flood forecasting, and river basin planning based on nationwide hydro-meteorological monitoring and remote sensing.

Forests and Biodiversity

Forest cover on a continuous, dramatic decline. Total forest cover in Cambodia declined from 13.1 million ha in 1973 to 8.4 million ha in 2018 (Figure 9 and Figure 10).[101] Dense old-growth forests, largely undisturbed and mostly evergreen, suffered the most losses, falling from 7.6 million ha in 1973 to 3.0 million ha in 2017. Official estimates indicate that the proportion of forest cover to total land area declined from 60% in 2006 to approximately 47% in 2018. Much of the remaining 8.4 million ha of forest cover has been degraded as a result of selective logging, unsustainable extraction of fuelwood, and land-use changes following indiscriminate awarding of economic land concessions.[102] Cambodia's forest cover has undergone significant changes over the past several decades. Along with the doubling of the population over the last 3 decades, land productivity for human settlement and agriculture substantially increased. Inadequate institutional capacity in terms of technical, personnel, and policy or planning dimensions in the face of massive human-induced development pressures (especially in low-lying and sensitive areas) and fundamentally weak governance arrangements have largely constrained the country's ability to improve the situation. Cambodia can learn from neighboring countries such as Viet Nam, where the government has succeeded in increasing forest cover against the total land as a result of long-term forest management since the 1980s (Figure 10).

---

[99] ADB and IHE Delft Institute for Water Education. 2019. *Characterizing Water Supply and Demand in Cambodia's River Basins*. Manila.

[100] The whole-of-system approach is one of the principles adopted by MOWRAM in the National Water Resources Management and Sustainable Irrigation Road Map and Investment Program, 2019–2033. It considers sustainable water resources management, complete irrigation schemes, and self-sustaining O&M as contributing to the goal of profitable irrigated agriculture, rather than the past approach that focused only on infrastructure development and rehabilitation.

[101] The percentage of nonforest ground cover (48.4%) has, for the first time, overtaken the share of forest cover (47.7%).

[102] Degradation in the upland areas, which are used for cassava cultivation, has been particularly high. Since 2012, a moratorium on new economic land concessions has been imposed. Forest Trends. 2015. *Conversion Timber, Forest Monitoring, and Land-Use Governance in Cambodia*. Washington, DC.

Figure 9: Total Forest Land Area, 2000–2018
('000 ha)

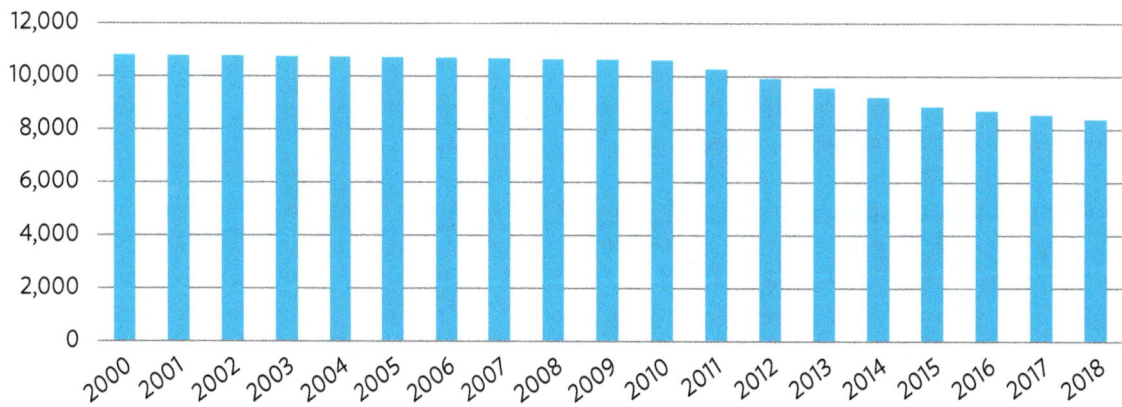

ha = hectare.
Source: Food and Agriculture Organization of the United Nations. FAOSTAT Database: Land Use (2000–2018) (accessed 7 June 2021).

Figure 10: Proportion of Forest to Total Land Area, 2000–2018
(%)

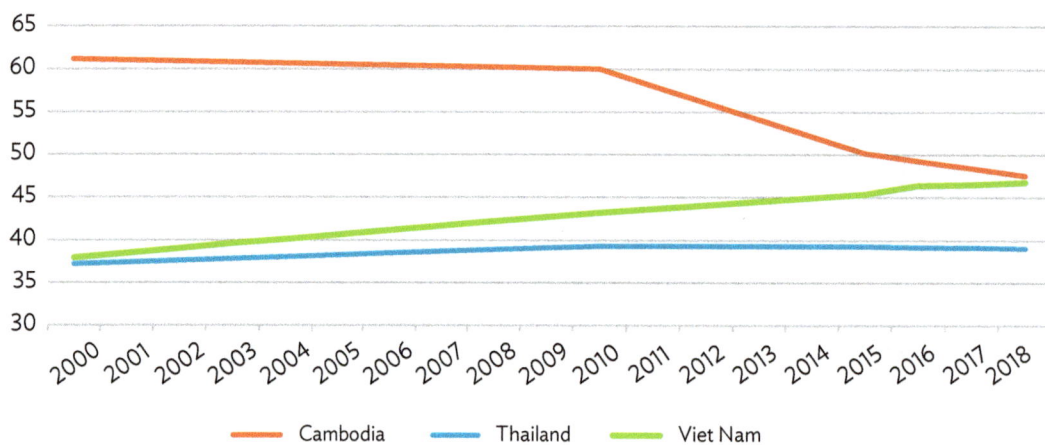

Source: Food and Agriculture Organization of the United Nations. FAOSTAT Database: Land Use (2000–2018) (accessed 7 June 2021).

Increasing forest cover in Cambodia. The Asian Development Bank has helped establish tree nurseries to increase forest cover in Cambodia (photo by Asian Development Bank Cambodia Resident Mission).

Strategies and programs for forest management. In support of achieving more sustainable forestry practices under Cambodia's Sustainable Development Goal target of 50% of national area under forest,[103] the MAFF established and approved 610 community forestry areas by the end of 2017.[104] The government has announced the application of a conservation corridor approach for the management of the country's forest landscapes. It has also developed a National Forest Programme, National Protected Area Strategic Management Plan, and National REDD+ Strategy. A major forest investment plan was prepared by ADB and the World Bank, which includes proposed investments in the main areas of (i) climate-smart landscapes and conservation corridors, (ii) reforestation and production forests through public–private partnerships, and (iii) implementing national forest monitoring.[105]

Rich biodiversity. Cambodia is home to a large variety of forests—from evergreen, coniferous, and deciduous forests, to mixed, flooded, and mangrove forests—covering almost half the country's land area. These forests are mainly found toward the northern and eastern regions of the country and in the southwest areas where population densities are lower. Cambodia's forests as well as coastal and riverine ecosystems provide habitats for diverse species of plants and wildlife. According to the Fifth National Report to the Convention on Biological Diversity, Cambodia is home to around 6,500 native flora and fauna species, including some endangered species such as the Asian elephant, tigers, giant catfish, the Irrawaddy dolphin, and numerous large water birds.[106] The country's forests serve as ecological buffers to natural disasters. They protect watersheds; act as carbon

---

[103] Government of Cambodia, Ministry of Environment, National Council for Sustainable Development (NCSD). 2016. *National Biodiversity Strategy and Action Plan*. Phnom Penh. p. 165.

[104] Footnote 8. These are established in accordance with the Forestry Law of 2002 and the Subdecree on Community Forestry of 2003.

[105] ADB and World Bank. 2017. Forest Investment Program: Cambodia's Investment Plan. Draft for review. 6 April.

[106] Government of Cambodia, Ministry of Environment, National Biodiversity Steering Committee. 2014. *Fifth National Report to the Convention on Biological Diversity*. Phnom Penh.

sinks; reduce soil erosion, thereby averting soil fertility loss; and prevent flooding, thereby protecting croplands in lower areas and slowing the sedimentation of reservoirs. Forests are particularly sensitive to the expected adverse impacts of climate change.[107]

Crosscutting Issues

Food Security, Safety, and Nutrition

Improving food security situation but persistent malnutrition. According to the 2019 Global Hunger Index, Cambodia's food security situation has improved substantially in recent years due to real income growth and volume increases in rice and other crop production.[108] However, this progress is now under threat due to the emergence of the COVID-19 pandemic. Comprehensive analysis is ongoing, but it appears the problem is more on the demand side (loss of income) than the supply side, and consumers are likely to be unable to afford sufficient and diverse food (footnote 12). While aggregate calorific availability at the national level is no longer an issue, stunting (low height-to-age ratio) is prevalent in households with poor sanitation. Also, localized or seasonal food deficits still occur and, when they do, the poorest households suffer the most. These households are usually landless, female-headed, with disabled family members, or households of ethnic minorities and/or living in the most remote and marginalized areas. There is a need to diversify national sources of nutrition and improve overall food quality for consumers. In 2018, some 70% of households reported medium dietary diversity and, partly in consequence, micronutrient malnutrition remains a major challenge.

Urgent need for food safety control improvement. In 2018, 12 people died and 203 people were hospitalized due to food poisoning.[109] The problems affecting food safety status can be summarized as follows: (i) the lack of a food safety policy and technical standards on food safety and management; (ii) no clear distinction in function on food safety control between the government and the private sector, and resultant underinvestment of the private sector; (iii) poor interministerial coordination and overlapping responsibilities; (iv) weak or ineffective border controls for inspection and monitoring; (v) inadequate staff capacity in food safety administration; (vi) insufficient trained manpower (e.g., for using laboratory equipment and testing kits); and (vii) lack of national laboratory facilities.[110] The absence of consumer representation or consumer voice limits the pressure for improvement. Having appropriate food safety control measures in place becomes even more critical in light of the current pandemic. To date, international trade requirements tend to drive the food safety agenda through global Hazard Analysis and Critical Control Points, the International Organization for Standardization (ISO), and sanitary and phytosanitary standards. Source ISO certifications are higher for rice than for aquaculture or horticulture, as food safety standards remain almost nonexistent. Pending the passage of the Food Safety Law,

---

[107] Government of Cambodia. 2015. Cambodia's Intended Nationally Determined Contribution to the United Nations Framework Convention on Climate Change. Phnom Penh. By 2050, it is projected that over 4 million ha of lowland forests, with a current dry season lasting between 4–6 months, will become vulnerable to water deficit periods of between 6–8 months or more. Lowland forests in the northeast and southwest are the most vulnerable.

[108] Concern Worldwide and Welthungerhilfe. 2019. *Global Hunger Index 2019: The Challenge of Hunger and Climate Change.* Dublin/Bonn.

[109] Foodborne pathogens can cause severe diarrhea and acute infections, such as meningitis. Chemically contaminated food can result in serious poisoning and life-threatening diseases, such as cancer and long-term disability. The increasing scale of intensified animal production calls for strengthening veterinary services, the control of animal diseases, food hygiene, and food safety standards to better protect public health. *Khmer Times.* 2019. Ministry Notes Decrease in Food Poisoning Cases. 11 February.

[110] WHO classifies food as "safe" when it is suitable for consumption and does not bring harm to the consumer when prepared and eaten according to its intended use. Controlling potential food-related hazards involves the application of control procedures and practices at all stages of the food chain, from primary production to processing and consumption.

an interministerial *prakas* (regulation or proclamation) forms the only framework for food safety in Cambodia.[111] It indicates accountability and the roles and responsibilities of ministries (e.g., MAFF, Ministry of Commerce, Ministry of Health) and authorized experts in standards, technical regulation development, legal framework, and policy development, among other points. However, coverage of the *prakas* is limited to food for commercial purposes only and the related food chain activities from the primary production stage at the farm level to the final consumption stage. The *prakas* excludes food used for direct household consumption and food for recreational purposes, animal feeding, and living modified organisms.

Climate Change, Disaster Risk Management, and Environmental Sustainability

Cambodia at high risk of climate change impacts. The country is presently ranked 12th out of 181 countries in the Global Climate Risk Index (over the period 1999–2018)[112] and 17th out of 180 countries in the World Risk Index in 2019.[113] In 2014, Standard & Poor's rated Cambodia's economy as the most vulnerable to the effects of climate change worldwide.[114] Similarly, the Notre Dame Global Adaptation Index presently assigns *high-vulnerability* and *low-readiness* scores to Cambodia.[115] In 2014, Cambodia's domestic vulnerability assessments indicated that 17.2% of Cambodia's communes (279 communes) were rated *highly vulnerable* and 31.5% (512 communes) were categorized *quite vulnerable* to multiple climate hazards.[116] Projections suggest that average rainfall will increase and become more concentrated, causing dry periods during the wet season and leading to more flood events; mean temperatures will rise, and drought periods will be extended.[117] A recent study reports that 10% of GDP could be lost by 2050 without adaptation to climate change.[118]

Climate adaptation policies and strategies developed. Cambodia's policy response to climate change has been positive. Line ministries have prepared sector climate change strategic plans supported by sound climate change action plans.[119] A national adaptation plan process has been initiated by the government that aims at strengthening ongoing climate adaptation policy responses through cross-sector programming, financing, and implementation.[120] The Cambodia Climate Change Alliance program is designed to strengthen the capacity of the National Council for Sustainable Development to serve its mandate of addressing climate change and enabling line ministries and civil society organizations to implement priority climate actions. The Strategic Program for Climate Resilience under the Climate Investment Funds emphasizes two streams to promote climate resilience: (i) developing knowledge of climate impacts in Cambodia and mainstreaming climate risk management into agriculture, water resources, transport, and urban infrastructure planning and management at a range of levels;

[111]   Government of Cambodia. 2010. *Inter-Ministerial Prakas on the Implementation and Institutional Arrangements of Food Safety Based on the Farm-to-Table Approach* (IMP868). *Prakas* No. UATH.BRK 868. Phnom Penh. 22 October.

[112]   D. Eckstein, V. Künzel, L. Schäfer, and M. Winges. 2019. *Global Climate Risk Index 2020*. Bonn: Germanwatch.

[113]   S. J. Day, T. Forster, J. Himmelsbach, L. Korte, P. Mucke, K. Radtke, P. Thielborger, and D. Weller. 2019. *World Risk Report 2019*. Berlin: Bündnis Entwicklung Hilft / Bochum: Ruhr University Bochum–Institute for International Law of Peace and Armed Conflict.

[114]   M. Kraemer. 2014. Climate Change is a Global Mega-Trend for Sovereign Risk. *Standard & Poor's Financial Services*. 15 May.

[115]   University of Notre Dame. Notre Dame Global Adaptation Initiative (ND-GAIN) Country Index (accessed 24 April 2020).

[116]   Government of Cambodia, Ministry of Environment, Department of Climate Change, General Secretariat of the NCSD. 2016. *Cambodia's National Climate Change Monitoring and Evaluation Framework*. Phnom Penh. April.

[117]   Since 1960, the mean annual temperature in Cambodia reportedly increased by 0.8°C. Also, the frequency of hot days and hot nights has increased significantly in every season. The average number of hot days per year has increased by 46 days. Since 1987, the frequency of extreme weather events (i.e., floods and droughts) has increased.

[118]   Government of Cambodia, Ministry of Economy and Finance, and NCSD. 2018. *Modelling of Climate Change Impacts on Growth*. Phnom Penh.

[119]   Fifteen ministries have developed climate change action plans, encompassing a total of 171 climate actions (7% are mitigation-oriented and 93% have an adaptation focus).

[120]   Government of Cambodia, Ministry of Environment, Department of Climate Change, General Secretariat of the NCSD. 2017. *National Adaptation Plan Process in Cambodia*. Phnom Penh.

and (ii) applying new skills, techniques, technologies, and engineering practices to climate-proof physical investments in a number of agriculture, water resources, urban, and infrastructure projects.[121]

There are two types of flooding in the country. First, floods resulting from overflows of the Mekong River, its tributaries, and the Tonle Sap Lake contribute to soil sedimentation and fertility, but also encroach on urban areas and seal soil. This flooding is slow to occur but lasts longer and can cause extensive damage. Second, heavy rains in mountainous areas can trigger flash floods in tributaries of the Mekong River and in streams. These floods can destroy crops and infrastructure in and along the tributaries of the Tonle Sap Lake. Drought is the second major hazard. It seriously impacts agricultural productivity, especially among rice farmers who depend on rain or rely solely on river-fed irrigation. Forest fires may also become increasingly common if forest degradation remains unchecked, thereby creating favorable conditions for large-scale forest fires.

Cambodia's agriculture sector is annually affected by crop damages from pests and by livestock and poultry losses from diseases, resulting in lower food production and loss of farmers' income. Swine flu and avian influenza have been long-standing concerns in Cambodia, given its rural landscape and the high exposure rates of farm animals.

Institutionalization of disaster risk reduction. Cambodia's Strategic National Action Plan for Disaster Risk Reduction (SNAP-DRR), 2008–2013 shifted the national paradigm from disaster response to disaster risk reduction (DRR).[122] An institutional structure on DRR was created within Cambodia, with the National Committee for Disaster Management at the national level and similar structures at subnational levels. Building on the SNAP-DRR, the National Action Plan for Disaster Risk Reduction was developed for 2014–2018.[123] Technical capacity, risk assessment, and monitoring remain weak, except for flood and storm or typhoon early warning systems. Community-based disaster risk management has gained traction in the SNAP-DRR period largely because of NGO involvement and support at the grassroots level.

High pressure on natural capital. The country's land-based ecosystems and natural capital are under major pressure from continuing rapid economic development, including agricultural activities. Expansion of the cultivated area has been a major contributing factor in the growth of agricultural production but has, at least partly, come at the expense of forests and wetlands. Human-induced activities such as unsustainable agricultural practices and slash-and-burn land clearance, combined with natural factors such as heavy rainfall, have further degraded Cambodia's already poor soil fertility. The pristine mangroves and coral reefs found along a 440 km coastline have also been under pressure from overexploitation of marine fishing in the past and from uncontrolled tourism and recreational developments. Cambodia's share of land degradation hotspots is among the largest globally, and about 60% of its population reside in these hotspot areas.[124] Because of its low level of environmental health and ecosystem vitality, the 2018 Environmental Performance Index ranked Cambodia 150th out of 180 countries.[125]

---

[121] Government of Cambodia, General Secretariat of the NCSD. Strategic Program for Climate Resilience–Cambodia. Financial support is provided from the Climate Investment Funds through ADB.

[122] Government of Cambodia, Ministry of Planning, and National Committee for Disaster Management. 2008. *Strategic National Action Plan for Disaster Risk Reduction, 2008–2013*. Phnom Penh.

[123] Government of Cambodia. 2013. *National Action Plan for Disaster Risk Reduction (NAP-DRR), 2014–2018*. Phnom Penh. November.

[124] Global information on the changes in the normalized difference vegetation index for 1982–2006 serves as a proxy for land degradation in the identification of these hotspots.

[125] Yale Center for Environmental Law & Policy, Yale University; and Center for International Earth Science Information Network, Columbia University. 2018. *2018 Environmental Performance Index*. New Haven.

Floods in rural areas. Flooding has caused severe damage to Cambodia's crops and livestock as well as to rural infrastructure (photo by Asian Development Bank).

Governance and Institutional Capacity

Weak governance. Cambodia's broad political-economic context remains problematic. The country continues to have the highest incidence of corruption among its structural peers and economic competitors and the highest bribery incidence in Southeast Asia.[126] Informal competition also disadvantages firms based in Cambodia.[127] Progress of reforms in public administration, public financial management, decentralization, and legal and judicial structure has been sluggish.[128]

Unclear institutional arrangements and incomplete legal frameworks. There are numerous overlaps in agency mandates, especially in the food safety domain where the Food Safety Law is not yet ratified and the respective responsibilities of the public and private sectors are not delineated. In the agro-processing sector, departments in both the MAFF and the Ministry of Industry and Handicraft have similar responsibilities and plans, which are not well-coordinated. There are also conflicts and tensions between the operations of major ministries. For example,

---

[126] In the 2019 Corruption Perceptions Index, Cambodia ranked 162nd out of 180 countries, with only the Democratic People's Republic of Korea (172nd) and Afghanistan (173rd) performing worse in the Asia-Pacific region. Transparency International. Corruption Perceptions Index 2019.

[127] World Bank. 2016. *Cambodia Enterprise Surveys*. Washington, DC. According to the Enterprise Surveys cited in World Bank's Systematic Country Diagnostic (footnote 23), there has been growing concern about the proliferation of informal business practices in Cambodia. In 2016, for example, 78% of firms declared they were competing against unregistered or informal firms, and this was highest among Cambodia's structural peers. Likewise, trust levels among value chain actors (e.g., between farmers and traders, or between farmers and input sellers) are very low.

[128] The World Justice Project's 2016 Rule of Law Index ranked Cambodia 112th out of 113 countries. K. Un. 2005. Patronage Politics and Hybrid Democracy: Political Change in Cambodia, 1993–2003. *Asian Perspective*. 29 (2). pp. 203–230.

the MAFF is responsible for water resources for on-farm irrigated agriculture and for catchment management programs, and the Ministry of Water Resources and Meteorology (MOWRAM) is responsible for the rehabilitation and construction of irrigation and water resources infrastructure, scheme O&M, and institutional issues related to the development and maintenance of irrigation infrastructure. MOWRAM's planning and investment programming need to adhere to national crop production policy and targets.

While there are numerous plans, strategies, and frameworks for the ANRRD sector, the emphasis is typically on broad development outcomes, with many action plans typically having limited connections to one another or lacking adequate resource allocation. The 2019 World Bank Public Expenditure Review of Cambodia describes how the introduction of program budgeting in major ministries (including the MAFF and MOWRAM) has improved the general alignment of budget expenditures with strategic goals and defined sector programs.[129] The review identified areas where key mandates are not adequately addressed (e.g., in relation to aspects of water management) and where allocations of resources across programs within sector strategies do not correspond to their anticipated relative importance. It acknowledges that the overall level of government resource allocation to the ANRRD sector is broadly appropriate to a country of middle-income status. The public expenditure review highlighted how future spending will need to be better targeted toward different areas, including agricultural research and sanitary and phytosanitary standards, and must become more strategic in irrigation investment.

Gender in Agriculture

Equal distribution of agricultural labor but highly unequal ownership of assets. Men are usually involved in land preparation and maintenance activities, including water and pest control. Women manage responsibilities that span pre- and postharvest activities. Women usually prepare and plant seeds, maintain seed beds, harvest and transport crops, and implement low-technology pest control measures by planting repellent grasses. Vegetables grown for subsistence are tended by women, while vegetables bound for market are tended by men; women are responsible for selling vegetables at the market. Small livestock operations sometimes supplement household incomes and diets. Women are traditionally responsible for pigs and poultry, while men care for larger animals like buffalo and cattle. Men typically catch fish, and women clean and sell them at the market.[130] Although women's participation in agriculture rose throughout the commodity boom, employment in garment, tourism, and service activities increasingly compete for women's rural labor, as do jobs in construction for men. Both lead to increased rural–urban migration. However, agriculture still accounts for over half of all women's employment, and women's participation rates in agriculture are falling more slowly than men's—in other words, the feminization trend of agriculture continues (footnote 130). Women now represent 74% of the agricultural workforce and produce 80% of Cambodia's food.[131] However, women own only 15.4% of the recorded agricultural land area (according to the 2017 Cambodia Socio-Economic Survey).[132] Women are estimated to receive only about 10% of all agricultural extension services, and female-headed households, on average, have less land and less access to farm equipment, tools, and communications services.

---

[129]   World Bank. 2019. *Improving the Effectiveness of Public Finance: Cambodia Public Expenditure Review*. Washington, DC.

[130]   ADB. 2015. *Promoting Women's Economic Empowerment in Cambodia*. Manila.

[131]   ADB. 2013. *Gender Equality in the Labor Market in Cambodia*. Manila.

[132]   Government of Cambodia, Ministry of Planning, National Institute of Statistics. 2018. *Cambodia Socio-Economic Survey 2017*. Phnom Penh. Table 1 and Figure 1. pp. 21–22.

Handbaskets by local women. The traditional Cambodian craft of bamboo weaving is predominantly carried out by women in the villages (photo by Asian Development Bank).

Regional Cooperation and Integration

Well-connected to the Greater Mekong Subregion. Cambodia's location in the Greater Mekong Subregion (GMS) offers major advantages, including access to over 300 million consumers and an increasingly affluent market for the ANRRD products on its doorstep. Intra-GMS trade continues to expand massively, at annual rates of over 16%, to a value exceeding $483 billion in 2017. Cambodia has benefited from its involvement in the GMS program since 1992, including through the GMS Core Agriculture Support Program and Core Environment Program, as well as from regional projects in transport and economic corridor development and in trade facilitation (e.g., customs procedures, sanitary and phytosanitary standards, and software investments).[133] More widely, the potential for Cambodia to improve its trade balance and deepen its own intersector and rural–urban linkages by intensifying value addition is enormous. Thanks to the bilateral trade agreement, Cambodia exported nearly 1 million tons of milled rice to the PRC by the third quarter of 2020, which comfortably offset the reduced import of milled rice by the European Union due to the end of the preferential treatment for imported goods from Cambodia.[134] However, Cambodia could further improve its trade balance if it can convert unofficial exports of unprocessed rice and other agricultural commodities (including fish and livestock) to neighboring countries to official exports of those commodities after processing.

---

[133] For Cambodia, the most important corridor is the Southern Economic Corridor (SEC), with routes linking Bangkok and Ho Chi Minh City via Phnom Penh.

[134] *Khmer Times*. 2020. Nearly 1 Million Tonnes of Milled Rice Exported to China by Q3. 30 October.

II. Sector Strategy

Government Strategy, Policy, and Plans

Cambodia's National Strategic Development Plan, 2019–2023. The National Strategic Development Plan was formulated for the implementation of the Rectangular Strategy.[135] The Rectangular Strategy is the highest level of policy and planning within which the ANRRD objectives may be located.[136] The promotion of the ANRRD sector looms large as one of the key priorities of the plan. With a target of 5% annual ANRRD sector growth, sector development objectives are stated in terms of increased productivity, crop diversification, and commercialization. The new Rectangular Strategy (Phase IV) involves enhancing the infrastructure for agricultural research, supporting technology improvement (including seeds) as well as pre- and postharvest technology development, promoting good agricultural practices, strengthening a range of agricultural services (including agricultural extension and information provision), strengthening agricultural cooperatives and developing agribusiness networks, promoting mechanization, improving land use management, promoting higher technology, and advocating for more resilient livestock and fisheries practices.

To support the objectives of the National Strategic Development Plan, 2019–2023, the MAFF's Agricultural Sector Strategic Development Plan, 2019–2023 covers (i) enhancement of agricultural productivity, diversification, and commercialization; (ii) promotion of animal health and production; (iii) management of fisheries and development of aquaculture; (iv) management and development of sustainable forestry and wildlife resources; and (v) improvement of efficiency of support services and human resources development.[137] Various agriculture subsector planning documents exist, including a master plan for agriculture sector development toward 2030;[138] a strategic planning document for livestock development to 2024;[139] and a Strategic Planning Framework for Fisheries, 2010–2019.[140] The Policy Document on the Promotion of Paddy Production and Export

---

[135] Government of Cambodia, Ministry of Planning. 2019. *National Strategic Development Plan, 2019–2023*. Phnom Penh. The National Strategic Development Plan (NSDP), 2019–2023 provides a summary of the key achievements, lessons, and challenges during the implementation of the preceding NSDP, 2014–2018. It also presents the macroeconomic and monitoring and evaluation frameworks for NSDP, 2019–2023 as well as outlines the policies and priority actions of relevant ministries for 2019–2023.

[136] The Rectangular Strategy is the principal socioeconomic policy agenda of the Government of Cambodia. It covers four growth factors: (i) agricultural development, (ii) infrastructure rehabilitation and development, (iii) private sector development and employment creation, and (iv) capacity building and human resources development.

[137] Government of Cambodia, MAFF. 2019. *Agricultural Sector Strategic Development Plan, 2019–2023*. Phnom Penh. The MAFF's newest 5-year plan was launched in December 2019.

[138] F. Goletti and S. Sin. 2016. *Development of Master Plan for Crop Production in Cambodia by 2030*. Final report. Prepared for the MAFF. Tonle Sap Poverty Reduction and Smallholder Development Project. Phnom Penh. May.

[139] Government of Cambodia, MAFF. 2015. *Strategic Planning Framework for Livestock Development, 2015–2024*. Draft report. 8 June.

[140] Government of Cambodia, MAFF, Fisheries Administration. 2011. *Strategic Planning Framework for Fisheries, 2010–2019: Fishing for the Future*. Phnom Penh. The Strategic Planning Framework for Fisheries in Cambodia was updated for the period 2015–2024, after the adoption of the new NSDP.

of Milled Rice (the rice policy) aims to promote Cambodia as an internationally recognized exporter of milled rice by shifting from the informal export of paddy rice to the formal export of milled rice.[141]

MOWRAM's Strategic Development Plan for Water Resources and Meteorology, 2019–2023 sets out strategies and targets for water resources management in five areas: (i) management and development of water resources, (ii) management of flood and drought, (iii) protection and conservation of water resources, (iv) management of information on water resources and meteorology, and (v) improvement of administrative management and human resources development.[142] In 2019, MOWRAM developed the National Water Resources Management and Sustainable Irrigation Road Map and Investment Program, 2019–2033, which identifies priority activities to be implemented over the next 15 years. With the aim of achieving a whole-of-system approach in water resources management and irrigation development, the program promotes (i) sustainable water resources management, (ii) complete irrigation schemes, (iii) self-sustaining O&M of irrigation systems, and (iv) profitable irrigated agriculture.[143]

Cambodia's Industrial Development Policy, 2015–2025. This policy aims to transform and modernize the industrial structure of Cambodia from a labor-intensive industry to a skill-driven industry by 2025.[144] This will be done by (i) connecting domestic industries to regional and global production value chains; (ii) developing interconnected production clusters and integrating into regional production networks, along with efforts to enhance productivity and strengthen the competitiveness of domestic industries; and (iii) moving toward developing a modern industrial base that is technology-driven and knowledge-based. The policy contains three specific targets: (i) to increase the GDP share of the industry sector to 30% by 2025 and drive the growth of the manufacturing sector to 20% by 2025 (from 15.5% in 2013); (ii) to diversify goods for export by increasing the share in total exports of non-textiles to 15% and of processed agricultural products to 12% by 2025; and (iii) to seek formal registration of 80% of small enterprises and 95% of medium enterprises, and to ensure that 50% of small enterprises and 70% of medium enterprises apply and maintain proper accounts and balance sheets. Priorities of the policy include supporting new industries, which can break into new markets with high-value, innovative, and competitive products; promoting small and medium-sized enterprises (SMEs); increasing production in the agriculture sector to serve both domestic and export markets; and encouraging the development of support industries for the agriculture, garment, and tourism sectors as they form part of cross-border production value chains.

Cambodia's Climate Change Strategic Plan, 2014–2023. The plan provides an overall structure for responding to and integrating climate change issues into development planning at national and sector levels.[145] It aims to increase resilience and promote low-carbon and green activities. Of the plan's eight strategic objectives, two are directly related to the ANRRD activities: (i) promoting climate resilience through improved food, water, and energy security; and (ii) supporting climate resilience of critical ecosystems (e.g., the Mekong River, the Tonle Sap

---

[141] Government of Cambodia, MAFF. 2013. *Policy Document on the Promotion of Paddy Rice Production and Export of Milled Rice.* Phnom Penh.

[142] Government of Cambodia, MOWRAM. 2018. *Strategic Development Plan for Water Resources and Meteorology, 2019–2023.* Phnom Penh. MOWRAM's strategic development plan will help implement the NSDP, 2019–2023 and the Rectangular Strategy (Phase IV). The ministry's overall mandate remains ensuring the sustainability of water resources for urban and rural water supply, agriculture, fisheries, hydropower, transportation, and tourism.

[143] Government of Cambodia, MOWRAM. 2019. *National Water Resources Management and Sustainable Irrigation Road Map and Investment Program, 2019–2033.* Phnom Penh.

[144] Government of Cambodia, Ministry of Industry and Handicraft. 2015. *Cambodia Industrial Development Policy, 2015–2025: Market Orientation and Enabling Environment for Industrial Development.* Phnom Penh. Approved by the Council of Ministers on 6 March.

[145] Government of Cambodia, National Climate Change Committee. 2013. *Cambodia Climate Change Strategic Plan, 2014–2023.* Phnom Penh.

Lake, highlands environment, and coastal habitats), protected areas, and biodiversity. All ministries have prepared climate change action plans.[146]

Responses to COVID-19. The government has started work on a post-COVID-19 pandemic recovery strategy encompassing all sectors of the economy. For agriculture, the country plans to promote more localized food production. Envisioned activities include the provision of finance and machinery for fruit and vegetable production, the introduction of aromatic rice varieties, and the promotion of industrial crops such as cassava and cashew nuts, among other measures. For the short term, the government encourages laid-off workers to return to their home provinces and start small-scale farming or work on their family farms. For these returning migrants, the government is looking into providing alternative means of income and employment through investments in areas such as forest farming and agro-forestry. To address the lack of access to finance and credit, the government is disbursing $50 million through the Rural Development and Agricultural Bank to provide low-interest loans to farmers and SMEs.[147] This capital injection aims to widen the agri-food system and will support production, processing, and agribusiness. The government has established the Credit Guarantee Corporation of Cambodia to provide a credit guarantee scheme to companies and businesses in priority sectors, including those severely affected by the pandemic. Lastly, Cambodia is piloting programs to improve access to and expand its social safety nets (e.g., e-payment systems for social transfers).[148]

ADB Sector Support and Experience

ADB sector program to date. Cambodia's transition from a planned to a market-oriented agriculture sector was supported by ADB through two sector program loans and a range of ANRRD investments, including the Tonle Sap Initiative. Since 2000, ADB has provided to the ANRRD sector 16 loans, 6 grants, and 37 technical assistance initiatives, with a combined amount of $569.8 million. ADB has supported the sector in four key areas: (i) enhancing agricultural productivity (e.g., by improving water resources planning and investment with better data and information management systems, improving irrigation efficiency and sustainability, enhancing water productivity, investing in irrigation schemes that can support rice and nonrice crops, and making on-farm practices more efficient and more sustainable); (ii) promoting diversification (i.e., a greater orientation toward potentially much higher-value outputs, including noncommodity outputs capable of value-added transformation); (iii) expanding commercialization and connectivity via improved value chains and market links, better linking primary producers to factor and product markets, rural and urban areas, and country-to-regional demand; and (iv) supporting sustainable natural resources management (including mapping and monitoring of ecosystem functions, strengthening of and support for natural resource institutions and processes), mainstreaming climate

---

[146] From the ANRRD perspective, the most relevant climate change action plans are the following: (i) MAFF—which emphasizes scaling up climate-smart technologies, researching and adopting more resilient crop varieties and livestock and fisheries practices, improving postharvest facilities, increasing the forestry carbon stock (including REDD+), and across-the-board capacity building (Government of Cambodia, MAFF. 2014. *Climate Change Priorities Action Plan for Agriculture, Fisheries and Forestry Sector, 2014–2018.* Phnom Penh); (ii) MOWRAM—which includes improved hydrological management and early warning systems, innovative technologies in areas affected by heavy rainfall, rehabilitation of flood protection dikes, construction of dikes in marine areas, and promoting gender responsiveness through climate planning (Government of Cambodia, MOWRAM. 2014. *Climate Change Action Plan for Water Resources and Meteorology, 2014–2018.* Phnom Penh); and (iii) Ministry of Rural Development (MRD)—which involves mapping vulnerable infrastructure and developing adaptation options, awareness-raising for village communities and within rural development planning, climate-proofing the Mekong River islands' connectivity (roads and jetties), among others (Government of Cambodia, MRD. 2014. *Climate Change Action Plan for Rural Development Sector, 2014–2018.* Phnom Penh).

[147] World Bank. 2020. *Cambodia Economic Update: Cambodia in the Time of COVID-19.* Cambodia Country Office. Phnom Penh.

[148] FAO. 2020. *The Impact of COVID-19 on Food and Agriculture in Asia and the Pacific and FAO's Response.* Paper prepared for the Thirty-Fifth Session of the FAO Regional Conference for Asia and the Pacific. Zoom Virtual Event. 1–4 September.

change, and improving access to climate-related finance. Recently completed and ongoing projects are listed in Appendix 2 and section III.

Self-evaluation highlights. There has been progress in overall public financial management reforms and funding of agricultural services, including extension, and general improvement in sector management capacity. At the same time, the need to move away from relatively straightforward tasks (like increasing the supply of basic commodities) toward more complex tasks (like increasing the production of niche rice and other products, reforming rice subsector policies, and urgently addressing climate change and disaster risk management) has somewhat increased and complicated sector management tasks even further. It has proven difficult to implement some earlier recommendations to improve sector investment performance. Completion reports for several recent ANRRD sector projects describe the following experiences: (i) initiating institutional reforms alongside investments in physical irrigation infrastructure took longer than expected; (ii) similar issues regarding lack of expected reforms to the sector institutional context were encountered in livelihoods projects; (iii) upgrades to small-scale rural infrastructure (i.e., social infrastructure, irrigation facilities, rural roads), when combined with training in improved farming and livestock practices, can have major impacts on the access and incomes of poor communities; and (iv) where government commitment to project objectives is strong and administrative capacity is robust, sector institutional and policy change can be achieved.[149]

Lessons learned and better practices. Given this experience in the ANRRD sector, the main lessons and best practices for the future are as follows: (i) fewer but bigger projects encourage both government commitment and critical implementation mass as well as reduced management costs; (ii) invest in areas where there is maximum government commitment and enthusiasm in evolving initiatives rather than simply replicating existing projects in certain subsectors; and (iii) address the underlying systemic issues surrounding O&M by strengthening the capacity of farmer water user communities (FWUCs), particularly in terms of collecting irrigation service fees, managing O&M funds, and looking at innovative public–private partnership approaches.

Other Development Partner Support

Large funding share goes to water resources management and irrigation. The ANRRD sector received funding support from development partners totaling around $177.7 million in 2017 and $263.0 million in 2019. Nearly two-thirds of this support goes to water resources and irrigation, with the next largest amount going to sector policy and management. Data from the official development assistance database of Cambodia confirms that, in 2018, ADB was the largest source of assistance to the agriculture sector. Other significant partners are Australia, the European Union, the International Fund for Agricultural Development (IFAD), Japan, the PRC, and the United States. Donor coordination in the sector is largely managed through the Technical Working Group on Agriculture and Water (TWGAW) and its subsector groups (e.g., irrigation, animal health, and agricultural extension). The TWGAW is comprised of the MOWRAM and the MAFF staff and Cambodia's development partners and is co-chaired by the two ministries.[150]

---

[149] ADB. 2014. *Completion Report: Northwest Irrigation Sector Project in Cambodia.* Manila; ADB. 2012. *Completion Report: Tonle Sap Sustainable Livelihoods Project in Cambodia.* Manila; ADB. 2017. *Completion Report: Tonle Sap Lowlands Rural Development Project in Cambodia.* Manila; and ADB. 2018. *Completion Report: Emergency Food Assistance Project in Cambodia.* Manila. For more details, see Appendix 2 of this report for the success ratings of the ANRRD completed loan and grant projects (Table A2.1) and technical assistance projects (Table A2.2) since 2008.

[150] The TWGAW is one of the technical working groups established under the umbrella of the Government of Cambodia's Donor Coordination Committee. ADB's Cambodia Resident Mission is a member of the TWGAW.

Poultry farming. Poultry production is part of an ADB-supported project that helps smallholder farmers generate more income through value addition in Thbong Khmum Province, Cambodia (photo by Asian Development Bank).

In terms of development partners with the closest affinity to ADB in the ANRRD sector at present, the IFAD, Australia's Department of Foreign Affairs and Trade,[151] and the World Bank stand out. The IFAD has a historic and ongoing role as a cofinancier in numerous ANRRD projects—and may be especially relevant in the future country partnership strategy period as the scope of its proposed Sustainable Assets for Agriculture Markets, Business and Trade Project is similar to ADB's emerging strategic directions in the sector. The IFAD's experience in the ongoing Agriculture Services Programme for Innovation, Resilience and Extension is extremely relevant.[152] Australia's Department of Foreign Affairs and Trade cofinanced a technical assistance initiative under the Water Resources Management Sector Development Program, and its Cambodia Agricultural Value Chain Program has much experience in establishing and maintaining self-reliant FWUCs. The World Bank is already a partner with ADB in the Forest Investment Program and is involved at scale in the development of a National Irrigation and Water Resources Investment Program.

In terms of technical expertise, the United States Agency for International Development (USAID) has in-depth experience in value chain development under the Helping Address Rural Vulnerabilities and Ecosystem Stability (HARVEST) program; the Japan International Cooperation Agency has considerable experience in rural infrastructure, national logistics, and support to agricultural cooperatives and research (e.g., on rice seed development and invasive pests). The Agence Française de Développement (AFD) has been a strong supporter of the sector, and its Water Resources Management and Agricultural Transition for Cambodia project will have several activities complementary to ADB's pipeline projects, including support to a water resources data management center.[153]

---

[151]   The Australian Agency for International Development (AusAID), which is responsible for the delivery of foreign aid from the Australian government, ceased to operate on 1 November 2013. Its work has now been integrated into the Department of Foreign Affairs and Trade.

[152]   IFAD. 2014. *President's Report: Proposed Loan and Grant to the Kingdom of Cambodia for the Agriculture Services Programme for Innovation, Resilience and Extension (ASPIRE)*. Rome.

[153]   For details, see Table A2.3 in Appendix 2 of this report for a list of development partner assistance in the ANRRD sector.

ADB's Sector Forward Strategy and Program, 2020–2023

ADB's strategy and assistance program. ADB's sector strategy will prioritize three key areas: (i) enhancing agricultural productivity, (ii) strengthening agricultural value chains, and (iii) strengthening natural resources management and disaster resilience. These three areas complement each other to support sector transformation toward being more productive, value-additive, and resource-efficient.

Together with ongoing projects, five forthcoming projects in the pipeline for 2021–2022 will form interventions from ADB's sector forward strategy. They are (i) the Community-based Tourism COVID-19 Recovery Project, for approval in 2021; (ii) the Sustainable Coastal and Marine Fisheries Project, for approval in 2021; (iii) the GMS Cross-Border Livestock Health and Value Chains Improvement Project, for approval in 2022; (iv) the Water Resources Management Investment Program, first tranche of a multitranche financing facility (MFF), for approval in 2022; and (v) the Strengthening Natural Resources Management and Disaster Resilience Project, for approval in 2022. These projects will cover one or more of the three key areas.

The following are the main considerations that guide the detailed scope of the interventions in ADB's three key areas of the sector forward strategy and program.

(i) Under the "new normal," overall emphasis should be on creating greater value addition in domestic processing and encouraging agro-processed exports.

(ii) Consistent with the emerging demands of a middle-income economy, food safety issues should be urgently addressed.

(iii) There is a need to involve and strengthen the private sector in the ANRRD operations and to help reduce the dichotomy that presently exists between an open, business-friendly foreign direct investment and macroeconomic environment, on the one hand, and the capacity, financial, and institutional problems that SMEs and cooperatives still face, on the other.

(iv) It is important to improve on-farm productivity, especially through more effective use of water in irrigation schemes by adopting a whole-of-system approach and by enhancing the stock of physical off-farm infrastructure (e.g., markets, collection, and storage facilities). This is becoming especially important with the emergence and evolution of the COVID-19 pandemic.

(v) There is a clear recognition for the need to confront natural resource loss and degradation, including pollution and waste management, brought about by intensifying agricultural practices and increased industrialization.

(vi) Climate change and disaster risk management considerations should be embedded throughout all sector operations.

(vii) ADB's three key areas are interconnected and will be pursued either in a single project or in coordination among ADB and non-ADB projects to enhance synergy. In either case, close coordination will be ensured to maximize the impacts of ADB's operations within the ANRRD sector and beyond.

Enhancing agricultural productivity. As part of the government's post-COVID-19 recovery plan, increasing local food production is a priority area for the agriculture sector. Enhancing agricultural productivity involves addressing crucial gaps in agricultural production. ADB supports (i) research and development on improved seeds and other planting materials and breeds, (ii) promotion of good agricultural practices, (iii) improvement in postharvest management, (iv) improvement in agricultural waste management, (v) strengthening of agricultural cooperatives, and (vi) irrigation and water management. ADB also supports diversification in the crop subsector and the livestock and fishery subsectors for long-term labor productivity increases and sustainable development of the agriculture sector. ADB promotes good agricultural practices in all subsectors to ensure environmentally sustainable agricultural production. The upcoming Community-based Tourism COVID-19 Recovery Project, Sustainable Coastal and Marine Fisheries Project, and GMS Cross-Border Livestock Health and Value Chains Improvement Project will contribute to this key area.

For irrigation and water management projects, which constitute a substantial part of the ANRRD sector operations in Cambodia, a whole-of-system approach will be undertaken through water resources management and the Irrigation Sector Road Map and Investment Program, 2019–2033 (the investment program), whereby profitable agriculture is placed at the center and determines irrigation investment that secures reliable water supply for different types of crops and supports fully functional irrigation schemes and sustainable O&M. By placing profitable agriculture at the center, connection with the agricultural value chain stream of investment will be ensured. In addition to infrastructure, the investment program will focus on software at the river basin and system levels, such as establishing decision support systems; conducting river basin water resources assessment, soil profile assessment, and ecological assessment; developing asset management systems; and establishing national irrigation benchmarking. The investment program is being supported by ADB and was endorsed in 2019. The forthcoming Water Resources Management Investment Program (MFF, first tranche) will support the investment program.

Strengthening agricultural value chains. Cambodia's agriculture sector needs transformation. Given better job opportunities in other sectors and neighboring countries, Cambodian agriculture needs to offer comparable on-farm income opportunities and additional off-farm and nonfarm jobs through developing value chains for key agricultural commodities. The COVID-19 pandemic has shown the importance of reliable connectivity between input providers, farmers, and buyers. To facilitate the sector's transformation, ADB will support agricultural value chain improvement. This approach is for efficient agro-based value chains that promote competitiveness and for food and nonfood agricultural products that are grown, processed, stored, and delivered in a safe, environment-friendly, and resource-efficient manner. This is in line with the Strategy for Promoting Safe and Environment-Friendly Agro-Based Value Chains in the Greater Mekong Subregion and Siem Reap Action Plan, 2018–2022.[154] Hence, this approach will address critical problems in value chains, which will include investments in rural infrastructure, agribusiness, and, to a limited extent, on-farm interventions. Engagement of the private sector is the key to competitive value chains. ADB provides financing to viable private enterprises and agricultural cooperatives for their investments in postharvest and logistic facilities, which are critical for strengthening value addition of selected Cambodian agricultural products (e.g., storage and warehouses; primary and deeper processing workshops; equipment, vehicles, and facilities for cold chains; and logistics facilities). ADB will support this key area through the forthcoming Sustainable Coastal and Marine Fisheries Project and GMS Cross-Border Livestock Health and Value Chains Improvement Project.

Strengthening natural resources management and disaster resilience. The key sectors driving economic growth and supporting livelihoods in Cambodia (such as agriculture, energy, external trade, industry, mining, tourism, transportation, and urban development) are highly dependent on the country's natural resources (land, water, forests, and minerals). Sustainably managing, protecting, and conserving Cambodia's environment and natural resources are vital in creating enabling conditions for economic growth; ensuring food, energy, and water security; and reducing poverty.[155] The COVID-19 pandemic has highlighted the consequences on livelihoods and human health that can result from continued biodiversity loss and degradation of ecosystems. To improve natural resources management and disaster resilience, ADB will support investments that focus on improved land use, watershed, and water resources management in targeted landscapes in the Tonle Sap and Lower

---

154   ADB. 2018. *Strategy for Promoting Safe and Environment-Friendly Agro-Based Value Chains in the Greater Mekong Subregion and Siem Reap Action Plan, 2018–2022.* Manila.

155   Government of Cambodia. 2017. *National Environment Strategy and Action Plan, 2016–2023.* Phnom Penh.

Mekong basins. The integrity of these landscapes, especially their watersheds, has been recognized as crucial for the range of ecosystem services they provide including dry-season water availability for multiple sectors, the fertility of productive agricultural lands for food production, fisheries, and other biodiversity resources. Specific interventions will include (i) adoption of soil and water conservation practices and technologies for integrated agriculture and water resources management within the agro-ecological landscape; and (ii) development of innovative financing mechanisms for sustainable watershed management, including payments for ecosystem services and forest carbon markets. In the coastal zone, ADB will support the improvement of coastal and marine resources and enhance the resilience of coastal zones through improved coastal and marine fisheries management and restoration of mangrove ecosystems. Mariculture and aquaculture practices that enhance ecosystem integrity will be explored, along with innovative sustainable financing mechanisms that promote private sector investment in conservation. ADB will support this key area through the forthcoming Sustainable Coastal and Marine Fisheries Project, Water Resources Management Investment Program, and Strengthening Natural Resources Management and Disaster Resilience Project.

Crosscutting Themes

ADB's future strategy and program will incorporate several crosscutting themes: (i) governance and capacity building, (ii) cross-sector synergies, (iii) regional cooperation and integration, (iv) gender development, and (v) COVID-19 pandemic recovery.

Governance and Capacity Building

All areas of support will continue to include governance strengthening and capacity building. Recent reforms in land and forestry management arrangements suggest the potential for modern land use monitoring systems and innovative financing arrangements. Capacity building across the sector will involve not just government staff but also the ANRRD institutions at the field level, including cooperatives, FWUCs, and other farmer groups (e.g., savings and credit groups). The inclusion of support for climate mainstreaming across all sector operations also enhances the relevance and quality of policy making.

Cross-Sector Synergies

Support for rural road improvement continues the previous logic of promoting physical connectivity between rural and urban areas. It will also facilitate the movement of labor and transport of goods to better address the impacts of the COVID-19 pandemic. The intensification of support for the later stages of the value chain through physical asset upgrade and SME financial and technical support encourages connectivity between primary and secondary production, or between agriculture and agro-industry. Work to address food safety improves the connectivity between food production and food consumption, and between rural and urban areas.

Regional Cooperation and Integration

Cambodia's continuing participation in the GMS Core Agriculture Support Program and the Core Environment Program will ensure it benefits from specific subregional investments (e.g., transboundary biodiversity corridors, economic corridor developments). The country's inclusion in thematic ANRRD initiatives will enable it to learn from more developed partners (e.g., the Siem Reap Action Plan). Similarly, participation in subregional sanitary and phytosanitary standards improvements will help facilitate increased formal export market access to neighboring countries and more widely. As part of value chain enhancements, transboundary livestock disease control will be facilitated through border disease control zones.

Gender Development

Investing in agriculture value chains, natural resources management, and climate change and disaster risk management can deliver strong gender benefits. In the proposed pipeline, gender mainstreaming will continue to be emphasized to help the government achieve its gender targets. Specifically, ADB projects and programs will (i) involve both women and men in project design, implementation, and decision-making (e.g., in community groups for irrigation and other rural services provision management); and (ii) provide more credit, knowledge, and technology activities for women.

COVID-19 Pandemic Recovery

ADB pays special attention to mitigating the impacts of the COVID-19 pandemic on the agriculture sector to help the country on the road to economic recovery. In addition to its ongoing efforts to strengthen production and value chains, ADB will leverage international best practices to be included in its upcoming projects. Elements include (i) enhancing food traceability throughout the value chain; (ii) leveraging digital innovations (e.g., e-commerce, financial services, market platforms, smart farming technology); (iii) promoting food production for local markets; (iv) facilitating improved access to agricultural and feed inputs; and (v) reinforcing and modernizing agricultural extension services.

Linkages between ADB Sector Strategy and the Government

The strategy and program reflect government policy in terms of supporting pre- and postharvest technology development, promoting good agricultural practices, strengthening a range of agricultural services and cooperatives and developing agribusiness networks, promoting mechanization, improving land use management, and encouraging more resilient livestock and fisheries practices, among other points. The specific objectives of the MAFF sector planning in terms of enhancing agricultural productivity, diversification, and commercialization are aligned with the proposed ADB interventions for better land and water resources management, extension of climate-smart and other agricultural technologies, and upgrading of postharvest facilities. Similarly, the MOWRAM's recognition of the need to improve the overall effectiveness of irrigation nationwide is reflected in the proposed ADB interventions, especially by addressing systemic O&M issues more directly in the context of a whole-of-system approach, as well as by employing more strategic planning, a prioritized investment approach, and better data upon which this can be based (footnote 100). The objectives of the Industrial Development Policy, 2015–2025 are underpinned by the reorientation of the ADB strategy and program toward the upper end of the agriculture value chains and by the broader support for business development in the ANRRD sector, especially for agro-industry SMEs. Continuing and large-scale interventions for climate change and disaster risk reduction in the ADB strategy and program also support the country's needs and government priorities, as reflected in the national and sector climate change action plans.

Risks and Assumptions

Sources of risk to the strategy. The design of the ADB strategy assumes that subregional and regional trade opportunities will continue to expand even though (i) the world commodity outlook is not especially positive, and (ii) it may be wise for Cambodia to reorient its export strategy to high-growth markets. Risks to the strategy arise from these assumptions not holding true and from (i) institutional commitment to reform not being as manifest in changed legislation and practices as in government policies and plans, as has been observed in past periods and in sector and project evaluations; and (ii) uncertainties and knowledge gaps with respect to some technologies and practices (e.g., in climate-smart agriculture) being unproven, with some aspects of government policy still being unclear in operational terms (e.g., regarding how public–private partnership arrangements may actually be approved and governed). One other source of risk relates to the absorptive capacity of the government. Given the similar technical scope of ongoing projects involving the same government agencies, the ability of the MAFF to successfully implement all these projects may be questionable.[156]

An unprecedented global pandemic risk. The world is now facing a novel and unprecedented risk with profound economic and social consequences causing massive disruption to the way of life as we knew it—the COVID-19 global pandemic.[157] COVID-19 has already claimed many lives. In developing countries such as those in the Association of Southeast Asian Nations (ASEAN), the pandemic has had a heavy toll on economic activities, including in the agriculture sector where challenges in food security and impacts on the agri-food supply chain are mounting. A report by Food Industry Asia recommends that "food and its broader supply chain be recognized as essential, along with the protection of the food industry labor force, ensuring borders between countries and within remain open and that financial assistance is provided to the most vulnerable businesses and consumers to minimize the impact of COVID-19 on food security in the ASEAN region."[158]

Although the impact of the pandemic was not as substantial as in some other countries such as the Philippines, Cambodia did place restrictions on movement at the outset of the pandemic that made it harder for farmers to sell their products to collectors who were unable to travel to farms, and farmers lacked the needed transport to take their products to markets.[159] There were also fewer people going to markets to buy goods, hence, reduced fresh market demand. Farmers had difficulty accessing agricultural inputs such as seeds, fertilizers, and veterinary medicines, thereby affecting production. Food losses and wastage were issues due to transport route blockages and checkpoints within and outside Cambodia. These problems were aggravated by the country's limited cold supply chain solution. In general, with the pandemic slowing down economies, food security and access as well as nutrition were adversely affected by job losses, income reductions, border closures, trade restrictions, food price increases, and local food availability.[160]

---

[156] Examples include the IFAD's Sustainable Assets for Agriculture Markets, Business and Trade Project, the World Bank's Cambodia Agriculture Sector Diversification Project, USAID's HARVEST Project II, and ADB's Climate-Friendly Agribusiness Value Chains Sector Project.

[157] World Food Programme. COVID-19 pandemic.

[158] Food Industry Asia (FIA) and PricewaterhouseCoopers (PwC). 2020. Maintaining Food Resilience in a Time of Uncertainty: Understanding the Importance of Food Value Chains in ASEAN and How to Ensure Their Resilience during the COVID-19 Crisis. Singapore.

[159] Sok Sotha. 2020. COVID-19: What Cambodian Farmers Are Experiencing? *World Farmers' Organisation*. 2 April.

[160] FAO. 2020. Q&A: COVID-19 Pandemic – Impact on Food and Agriculture. Rome.

Investment and Technical Assistance Programs

Table 1 summarizes projects in the proposed ANRRD pipeline for 2021–2023.

Table 1: Agriculture, Natural Resources, and Rural Development Sector—Pipeline Projects, 2021–2023

Project Name	Loan/Grant Amount ($ million)	Loan/Grant Approval Year	Lead Agency
Community-Based Tourism COVID-19 Recovery Project	3.0[a]	2021 (Firm)	MAFF
Sustainable Coastal and Marine Fisheries Project (National)	55.0	2021 (Firm)	MAFF
Water Resources Management Investment Program, MFF, first tranche (National)	140.0	2022 (Firm)	MOWRAM
GMS Cross-Border Livestock Health and Value Chains Improvement Project	75.0	2022 (Firm)	MAFF
Strengthening Natural Resources Management and Disaster Resilience Project	82.0	2022 (Firm)	MAFF
Total	**355.0**		

GMS = Greater Mekong Subregion; MAFF = Ministry of Agriculture, Forestry and Fisheries; MFF = multitranche financing facility; MOWRAM = Ministry of Water Resources and Meteorology.

[a]   To be financed by a grant from the Japan Fund for Poverty Reduction.

Source: Asian Development Bank.

III. Sector Road Map and Results Framework

Table 2 highlights the road map and results framework of Cambodia's ANRRD sector for 2019–2023.

Table 2: Agriculture, Natural Resources, and Rural Development Sector—Road Map and Results Framework, 2019–2023

Country Sector Outcomes		Country Sector Outputs		ADB Sector Operations	
Sector Outcomes with ADB Contribution	Indicators with Targets and Baselines	Sector Outputs with ADB Contribution	Indicators with Incremental Targets	Planned and Ongoing ADB Interventions	Main Outputs Expected from ADB Interventions
Agricultural productivity, diversification, competitiveness, and commercialization increased (ASSDP)	Annual value added in agriculture increased to KR25.69 billion by 2023 (2019 baseline: KR22.61 billion) (NSDP) Proportion of crop diversification increased to 67.8% by 2023 (2019 baseline: 63.4%) (ASSDP) Industrial crop production increased to 25.5 million tons by 2023 (2019 baseline: 17.4 million tons) (ASSDP) Share of processed agricultural products export in total exports increased to 10% by 2023 and 12% by 2025 (2019 baseline: 7%) (NSDP/CIDP) Commercial livestock production as percentage	Agricultural productivity, diversification of crops, and agribusiness increased (ASSDP) Animal production and animal health promoted (ASSDP)	Annual value added of crop production increased by 3.1% and livestock production by 2.7% during 2019–2023 (ASMP) Rice production increased by 3.0% annually, to 12.6 million tons by 2023 (2019 baseline: 11.2 million tons) (ASSDP) Rice yields increased by 2.2% annually, to 3,780 tons/ha by 2023 (2019 baseline: 3,450 tons/ha) (ASSDP) Cultivated areas for key industrial crops[a] increased to 1.70 million ha by 2023 (2019 baseline: 1.57 million ha) (ASSDP) Exports of agricultural crops increased to 7.7 million tons by 2023	**Planned key support areas:** 1. Enhancing agricultural productivity 2. Strengthening agricultural value chains 3. Strengthening natural resources management and disaster resilience **Pipeline projects with estimated amounts:** Community-Based Tourism COVID-19 Recovery Project (2021: $3.0 million JFPR grant) GMS Cross-Border Livestock Health and Value Chains Improvement Project (2022: $50.0 million OCR concessional loan, $12 million ADF grant, $13 million cofinancing [TBC]) Sustainable Coastal and Marine Fisheries Project (2022: $50.0 million OCR concessional loan, $5 million ADF grant, $40 million cofinancing [TBC]) Strengthening Natural Resources Management and Disaster Resilience Project (2022: $80.0 million OCR concessional loan, $2 million ADF grant)	**Pipeline projects:** 20,000 ha of land improved through irrigation and drainage and flood management 80,505 persons with sustainable access to improved water source (rural) 10 SMEs loan accounts opened, and end borrowers reached (number of accounts) **Ongoing projects:** 3.3–3.5 tons/ha average rice yields increased; 4.25 tons/ha average annual crop production increased At least 214 agribusinesses or microenterprises are operational or established 117,203 ha of land improved through irrigation and drainage and flood management 1,277 km of rural roads built or upgraded 5,600 ha of natural forest land rehabilitated through forest restoration, enrichment planting, and non-timber forest product

continued on next page

Table 2 *continued*

Country Sector Outcomes		Country Sector Outputs		ADB Sector Operations	
Sector Outcomes with ADB Contribution	Indicators with Targets and Baselines	Sector Outputs with ADB Contribution	Indicators with Incremental Targets	Planned and Ongoing ADB Interventions	Main Outputs Expected from ADB Interventions
	of total animal production increased to 30% by 2023 (2019 baseline: 23%) (ASSDP)		(2019 baseline: 5.9 million tons) (ASSDP/NSDP) Commercial animal production increased to 16.1 million heads by 2023 (2019 baseline: 11 million heads) (ASSDP)	Integrated Water Resources Management Project (2023: $75.7 million OCR concessional loan, $4.3 million ADF grant, $60.0 million cofinancing [TBC]) **Ongoing projects with approved amounts:** GMS Flood and Drought Risk Management and Mitigation Project (2012: $35.0 million ADF concessional loan) Climate-Resilient Rice Commercialization Sector Development Project (2013: $55.0 million ADF concessional loan) GMS Biodiversity Conservation Corridors Project–Cambodia: Additional Financing (2015: $19.0 million ADF grant)	and agroforestry planting using mainly native species 2,000 households or 10,000 individuals with sustainable access to improved water source (rural) 965 households or 3,860 individuals living on less than $1 per day who will earn more than $1 per day after project 20,000 microfinance loan accounts opened, or end borrowers reached 40,000 tCO$_2$e per year of GHG emissions reduced
Sustainable management and development of land, forest, and fisheries resources optimized (ASSDP)	Community forestry revenue increased to 18% by 2023 (2019 baseline: 10%) (ASSDP) Rate of aquaculture compared to total fisheries products increased to 49% by 2023 (2019 baseline: 32%) (ASSDP)	Sustainable conservation, management, and development of forest, fisheries, and aquaculture resources strengthened (ASSDP/CCCSP)	Annual value added of forestry products increased by 1.2% and fisheries products by 4.3% during 2019–2023 (ASMP) Total annual fish catch (from natural and aquaculture) increased to 1,219,400 tons by 2023 (2019 baseline: 898,700 tons) (NSDP)	Uplands Irrigation and Water Resources Management Sector Project (2015: $60.0 million ADF concessional loan) Tonle Sap Poverty Reduction and Smallholder Development Project–Additional Financing (2017: $49.13 million OCR concessional loan, $31.57 million ADF grant) Climate-Friendly Agribusiness Value Chains Sector Project (2018: $90.0 million OCR concessional loan, $10 GCF loan, $30 million GCF grant)	
Infrastructure (e.g., irrigation, roads) and trade facilitation (e.g., credit, investment) in agriculture strengthened (ASMP)	Rate of irrigated rice crop capability (rainy and dry seasons) increased to 58.3% by 2023 (2019 baseline: 55.3%) (ASMP) Rural road infrastructure (rehabilitated and built) increased to 43,570 km in 2023	Effectiveness of irrigation enhanced (ASMP) Road connections and transportation of agricultural products expanded (ASMP) Agricultural investment and credit promoted (ASMP)	Total rice cultivated areas (dry and rainy seasons) with access to irrigation system increased to 1.952 million ha by 2023 (2019 baseline: 1.832 million ha) (NSDP/SDPWRM) Annual potential irrigated additive crops increased to	Irrigated Agriculture Improvement Project (2019: $117.0 million OCR concessional loan, $2.16 million ADF grant) Agricultural Value Chain Competitiveness and Safety Enhancement Project (2020: $70.0 million OCR concessional loan, $5.0 million AIF loan, $25.0 million AFD loan, $3.0 million JFPR grant)	

continued on next page

Table 2 *continued*

Country Sector Outcomes		Country Sector Outputs		ADB Sector Operations	
Sector Outcomes with ADB Contribution	Indicators with Targets and Baselines	Sector Outputs with ADB Contribution	Indicators with Incremental Targets	Planned and Ongoing ADB Interventions	Main Outputs Expected from ADB Interventions
	(2019 baseline: 30,740 km) (NSDP) Loans provided to the nonrice sector (agro-industrial crop production, animal raising, and aquaculture) increased to 42.0% by 2023 (2019 baseline: 23.3%) (ASMP)		6,913 ha by 2023 (2019 baseline: 4,913 ha) (NSDP) At least 5% of total length of rural roads connect to farms, to 2,190 km by 2023 (2019 baseline: 1,500 km) (ASMP) Investments in agro-industries increased to KR2,948 billion by 2023 (2019 baseline: KR2,012 billion) (ASMP) Credit for the agriculture sector and agro-industry from the ARDB increased to $682.5 million by 2025 (2019 baseline: $198.0 million) (ASMP)		
Climate resilience strengthened through improving food, water, and energy security; and by promoting low-carbon planning and technologies to support sustainable development (CCCSP)	Research for increasing agricultural productivity and quality developed to 11 innovative technologies by 2023 (2019 baseline: 8 technologies) (ASSDP) Proportion of population in rural areas with access to safe and clean water supply services increased to 90% by 2023 (2019 baseline: 65%) (NSDP) Percentage of climate-change-	Research for sustainable agriculture development enhanced (ASSDP) Capacity on new technology on crops, livestock production, forestry, and fisheries that have adaptive capability to the impacts of climate change[b] strengthened (CCCSP) GHG emissions from deforestation and forest degradation, livestock and crop	Agricultural production increased by 5.0% annually during 2019–2023 by using innovative technology (ASSDP) Cultivated area of weather-resistant rice crops increased to 1.32 ha by 2023 (2019 baseline: 1.2 ha) (NSDP) Length of climate-change-adapted rural roads increased to 2,251 km by 2023 (2019 baseline: 640 km) (NSDP)		

continued on next page

Table 2 *continued*

Country Sector Outcomes		Country Sector Outputs		ADB Sector Operations	
Sector Outcomes with ADB Contribution	Indicators with Targets and Baselines	Sector Outputs with ADB Contribution	Indicators with Incremental Targets	Planned and Ongoing ADB Interventions	Main Outputs Expected from ADB Interventions
	adapted national and provincial roads increased to 67.2% by 2023 (2019 baseline: 55.3%) (NSDP)	production, and primary production reduced[c] (ASSDP/CCCSP)	GHG emissions compared to the base year data reduced by 3,018,000 GgCO$_2$e by 2023 (2019 baseline: 2,799,000 GgCO$_2$e) (NSDP)		

$ = United States dollar, ADB = Asian Development Bank, ADF = Asian Development Fund, AFD = Agence Française de Développement, AIF = ASEAN Infrastructure Fund, ARDB = Agriculture and Rural Development Bank, ASEAN = Association of Southeast Asian Nations, ASMP = Agricultural Sector Master Plan, ASSDP = Agriculture Sector Strategic Development Plan, CCCSP = Cambodia Climate Change Strategic Plan, CIDP = Cambodia Industrial Development Policy, COVID-19 = coronavirus disease, GCF = Green Climate Fund, GgCO$_2$e = gigagram of carbon dioxide equivalent, GHG = greenhouse gas, GMS = Greater Mekong Subregion, ha = hectare, JFPR = Japan Fund for Poverty Reduction, km = kilometer, KR = riel, NSDP = National Strategic Development Plan, OCR = ordinary capital resources, SDPWRM = Strategic Development Plan on Water Resources and Meteorology, SMEs = small and medium-sized enterprises, TBC = to be confirmed, tCO$_2$e = ton of carbon dioxide equivalent.

[a]   Among Cambodia's key industrial crops are maize, cassava, cashew, banana, mango, pepper, and sugarcane.

[b]   These impacts include drought, flood, temperature rise, saline intrusion, and destruction from insects and diseases.

[c]   Measures include encouraging the management, conservation, and sustainable development of forest resources, especially community forestry, as well as promoting the use of renewable energy (biomass and biogas) and the appropriate use of agricultural technology.

Source: ADB.

Appendix 1

SWOT and Problem Tree Analyses of Cambodia's Agriculture, Natural Resources, and Rural Development Sector

Sector SWOT Analysis

Based on the description of the agriculture, natural resources, and rural development (ANRRD) subsectors and conditions, Table A1 summarizes the strengths, weaknesses, opportunities, and threats (SWOT) for the sector. The sector's **strengths** lie largely in its underlying natural resource base and human resource potential, coupled with what has so far been demonstrated in positive policy and management terms. **Weaknesses** arise out of the wider political economy of Cambodia, including generally low accountability and transparency, inadequate or incompletely applied legal arrangements for natural resources management, deficiencies in factor markets, and relatively limited output opportunities.

The **opportunities** for Cambodia's ANRRD sector are determined by its potential to supply its growing domestic food demand as well as the large regional market on its doorstep to which it is becoming increasingly better connected. The scope for widespread productivity improvements across all agricultural, livestock, and fisheries activities is enormous, as is the potential for more sustainable natural resources management. **Threats** to realizing these opportunities are both external (e.g., from competitor countries, changing export market arrangements, climate change, the Mekong mainstream dams) and internal (e.g., from migration trends, labor availability, commitment to reform).

Table A1: Agriculture, Natural Resources, and Rural Development Sector—SWOT Analysis

Strengths	Weaknesses
• Large and diverse national water resources	• Highly seasonal water resources availability; flat topography limiting irrigation potential and efficiency
• Relatively abundant land and soils suitable for a range of agricultural activities	• Systemic issues with finance for O&M and farmer irrigation management transfer
• Diversity of ecosystems (flatlands and hills, flooded forests, riverine and coastal zones)	• Poor ANRRD sector budgeting, planning, data, and implementation capacity; low accountability and transparency
• Relatively young and adaptable labor force	• Weak public sector management and oversight in the ANRRD area and wider services (e.g., extension, inputs regulation, food safety, disaster risk management)
• On-farm underlying comparative advantage in rice production (especially aromatics)	
• Stable macroeconomic and policy framework, large foreign direct investment flows, sustained economic growth over a long period	• Undeveloped legal instruments for land, agriculture, and water resources management
• Export-oriented national planning framework in place	• Heavy dependence on production and exports of a few unprocessed commodities (e.g., rice, rubber); lack of output diversification and limited value addition

continued on next page

Table A1 *continued*

Strengths	Weaknesses
• Increasing sophistication of the ANRRD sector planning (e.g., climate change mainstreaming; intersector linkages, especially agro-industry) • Some notable examples of successful subsector planning and administration initiatives (e.g., rice policy formulation, land titling, trade facilitation), and some components of investment projects • Varied and (now) widespread availability of farm-level agricultural credit • Long-term development partner support	• Technology, finance, information, and entrepreneurship constraints to agro-processing expansion • Generally limited access to and low use of improved crop technology (especially on small farms) and, consequently, low yields by regional standards • Weak intersector linkages and physical connectivity of primary producers to markets (domestic and international); high off-farm processing, transport, energy, and transaction costs • Limited milling capacity • Limited social and institutional capacity at all levels (e.g., farm and community, ANRRD business associations)

Opportunities	Threats
• Large and rapidly growing regional demand for all ANRRD products (commodities, fruits and vegetables, livestock, processed foods) • Increasing domestic demand for nutritious, safe, processed food products (especially in urban areas and from tourists) • Strong global demand for Cambodian aromatic rice varieties • Ongoing investments in value chains—physical infrastructure (special economic zones, agro-industry clusters); transport (corridors, roads, haulage services); software (e.g., Cross-Border Transportation Agreement)—provide better links from Cambodia to wider GMS, ASEAN, and global markets • Wide scope to raise crop and livestock productivity across farms of all sizes—increasing primary output, processing potential, and sector incomes • Contract farming potential (including inputs supply, integrated production models, cooperatives) • Major potential to apply new water and agricultural technologies (drip irrigation, water harvesting, biomass, rural energy) and mechanization • Potential for natural or green infrastructure in water resources management and development; community resources management	• Global demand and forecast prices for unprocessed commodities (especially rice) deteriorating (e.g., long-term decline in agricultural terms of trade); EU's "Everything but Arms" access arrangement compromised • ANRRD product range (including aromatic rice) competing with larger regional countries (Thailand, Viet Nam, Myanmar); Cambodia remaining as marginal global supplier • Domestic ANRRD processing competing with economies of scale available in neighboring countries • Quantity and quality of natural resource base (i.e., forest, soil, and water) under pressure (e.g., the Mekong dams, agrochemicals use, land sequestration) • Polarization of farm sizes (large farms becoming larger, small ones smaller) resulting in limited income growth potential, increased landlessness, lack of tenure, and informal land trade • Increasing competition and scarcity of rural labor; large-scale migration to urban areas (for employment in construction and garment industries) and other countries • Climate change impacts becoming increasingly severe (incidence of flood and drought rising, disasters increasing); underresourced disaster risk management needs • Slow progress with administrative, legal, and decentralization reforms • Impact of the COVID-19 pandemic and the associated economic downturn

ANRRD = agriculture, natural resources, and rural development; ASEAN = Association of Southeast Asian Nations; COVID-19 = coronavirus disease; EU = European Union; GMS = Greater Mekong Subregion; O&M = operation and maintenance; SWOT = strengths, weaknesses, opportunities, and threats.

Source: Asian Development Bank.

Problem Tree Analysis

The Problem Tree (Figure A1) summarizes the current issues in Cambodia's ANRRD sector in terms of cause and effect relationships. The core ANRRD sector problem is defined as low factor productivity, very limited value addition, and threatened natural capital stock vulnerable to climate-related shocks. This problem has three main causes.

First, **the ANRRD output is dominated by the production of largely unprocessed crop commodities**, especially rice, but also rubber, maize, and cassava. There is a lack of diversification in the product mix, with relatively few noncommodity outputs (e.g., fruits, vegetables, higher-value tree or industrial crops). In addition, the low level of mainly wet-season rice yields keeps overall sector productivity low despite Cambodia's comparative advantage in rice production.

Factors that contribute to this situation include (i) prevalence of low levels of technology in wet-season cropping, especially on smaller farms; (ii) modest level of irrigation impact, with limited success in achieving irrigation management transfer and expansion of dry-season and high-value cropped areas; (iii) weaknesses in public research and agricultural extension systems; (iv) unreliable and inadequate inputs access, including seeds, and limited farmer input application, knowledge, and control; (v) poor livestock and fisheries nutrition and management practices, with deficient input supplies and farmer knowledge; and (vi) inadequate local rice milling capacity and associated high milling costs.

Second, **Cambodia's agriculture and livestock value chains are underdeveloped.** This has physical and institutional causes, including (i) poor quality of rural roads and high cost of farm-to-market transportation; (ii) deficit of markets, stores, warehouses, cold stores, and larger agro-processing infrastructure facilities, including special economic zone and agro-processing clusters; (iii) a "missing middle" in the agribusiness distribution network—specifically, there are relatively few small and medium-sized enterprises (SMEs), and those that do exist have limited capacity (in terms of access to information, technical and business skills, and finance) and have to deal with a weak legislative investment framework; (iv) underdeveloped farmer and producer organizations (especially agricultural cooperatives for inputs and machinery supply and outputs marketing) and immature contract farming arrangements; and (v) inadequate food safety legislation and enforcement, and absence of safety and traceability mechanisms.

Third, **the ANRRD sector is characterized by unsustainable natural resources management practices, which are increasingly compounded by climate change impacts.** The factors contributing to this core sector problem include (i) weak, overlapping, and incomplete legal frameworks administered by the ANRRD institutions with limited technical capacity; (ii) weak control of (mostly informally imported) agricultural inputs and other pollutants—human, soil, water contamination, etc.; (iii) slow closure of the national climate adaptation deficit (i.e., climate-suitable or climate-smart technologies not fully applied, and climate-proofing rural infrastructure not occurring quickly enough); (iv) rising flood and drought risks and costs, with increasing climate variability; and (v) underdeveloped community resources and national and subnational disaster risk management systems.

The impact of the core sector problem **within the ANRRD sector** is seen in terms of (i) limited export earnings, continuation of processed food imports, and a prevalence of insecure or unsafe domestic food products; (ii) low farm incomes and limited rural nonfarm job opportunities creation; and (iii) continuing deterioration of land, forest, and water resources, and magnification of climate-related risks.

Figure A1: Agriculture, Natural Resources, and Rural Development Sector—Problem Tree

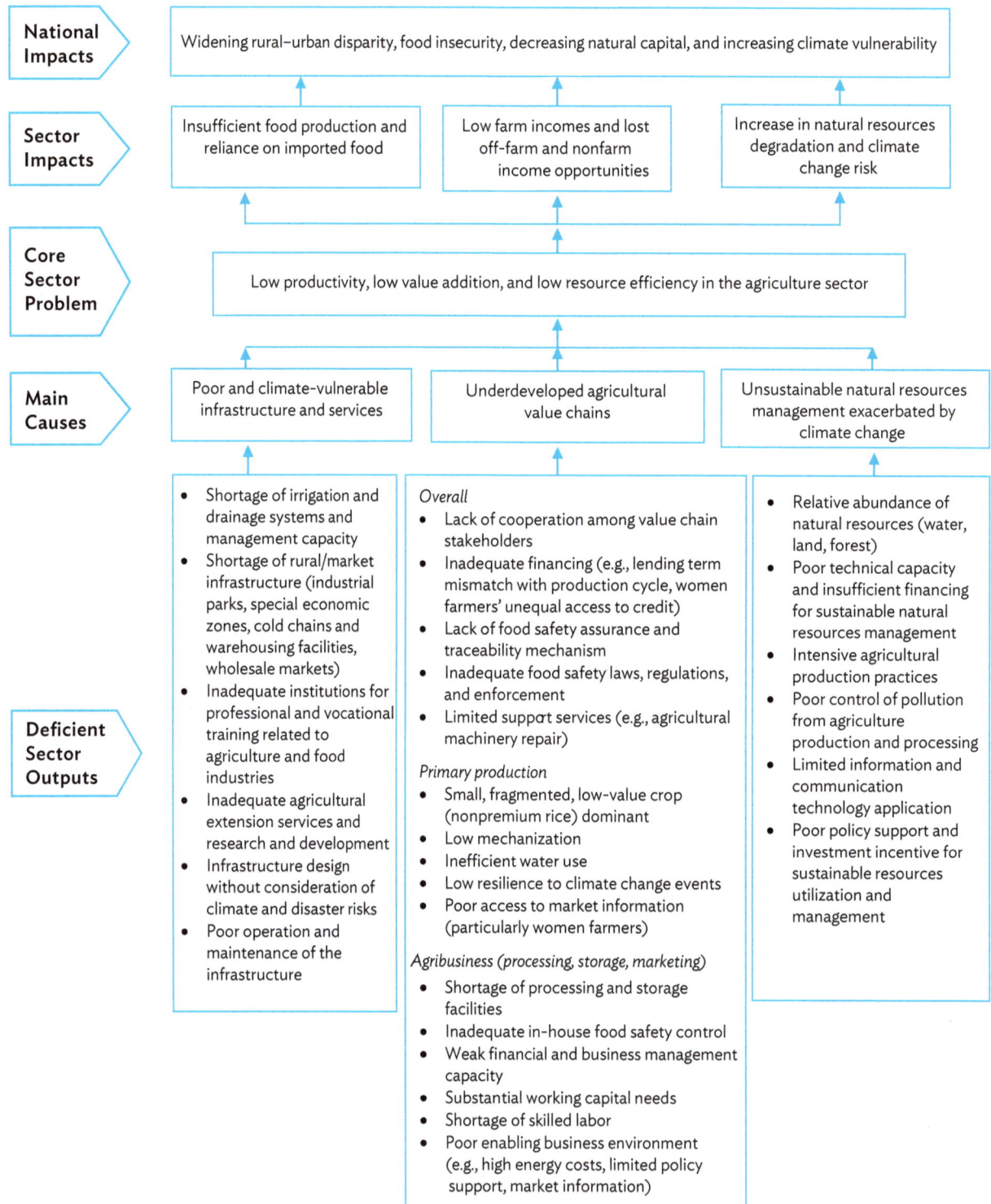

National Impacts

Widening rural–urban disparity, food insecurity, decreasing natural capital, and increasing climate vulnerability

Sector Impacts

| Insufficient food production and reliance on imported food | Low farm incomes and lost off-farm and nonfarm income opportunities | Increase in natural resources degradation and climate change risk |

Core Sector Problem

Low productivity, low value addition, and low resource efficiency in the agriculture sector

Main Causes

| Poor and climate-vulnerable infrastructure and services | Underdeveloped agricultural value chains | Unsustainable natural resources management exacerbated by climate change |

Deficient Sector Outputs

- Shortage of irrigation and drainage systems and management capacity
- Shortage of rural/market infrastructure (industrial parks, special economic zones, cold chains and warehousing facilities, wholesale markets)
- Inadequate institutions for professional and vocational training related to agriculture and food industries
- Inadequate agricultural extension services and research and development
- Infrastructure design without consideration of climate and disaster risks
- Poor operation and maintenance of the infrastructure

Overall
- Lack of cooperation among value chain stakeholders
- Inadequate financing (e.g., lending term mismatch with production cycle, women farmers' unequal access to credit)
- Lack of food safety assurance and traceability mechanism
- Inadequate food safety laws, regulations, and enforcement
- Limited support services (e.g., agricultural machinery repair)

Primary production
- Small, fragmented, low-value crop (nonpremium rice) dominant
- Low mechanization
- Inefficient water use
- Low resilience to climate change events
- Poor access to market information (particularly women farmers)

Agribusiness (processing, storage, marketing)
- Shortage of processing and storage facilities
- Inadequate in-house food safety control
- Weak financial and business management capacity
- Substantial working capital needs
- Shortage of skilled labor
- Poor enabling business environment (e.g., high energy costs, limited policy support, market information)

- Relative abundance of natural resources (water, land, forest)
- Poor technical capacity and insufficient financing for sustainable natural resources management
- Intensive agricultural production practices
- Poor control of pollution from agriculture production and processing
- Limited information and communication technology application
- Poor policy support and investment incentive for sustainable resources utilization and management

Source: Asian Development Bank.

At the national level, the impacts of the ANRRD conditions are felt in terms of (i) low and variable contributions to national growth, (ii) inadequate national dietary support, (iii) continuing migration (internal and to other countries), and (iv) insufficiently reduced vulnerability to climate change and natural disaster.

This **assessment of the ANRRD sector is in line with the strategic development plan of the Ministry of Agriculture, Forestry and Fisheries,**[1] which emphasizes the need to (i) improve overall productivity, including in rice production, but also more widely; (ii) invest in postharvest infrastructure; (iii) improve the regulatory and enforcement environment, including with respect to food safety and control of animal diseases; (iv) strengthen the technical and market access capacity of SMEs and farmer organizations; (v) increase the role of the private sector; (vi) improve forest, land, and biodiversity management; and (vii) urgently address climate-related needs.

---

[1]   Government of Cambodia, Ministry of Agriculture, Forestry and Fisheries. 2015. *Agricultural Sector Strategic Development Plan, 2014–2018.* Phnom Penh.

Appendix 2
Projects by ADB and Other Major Development Partners in Cambodia's Agriculture, Natural Resources, and Rural Development Sector

Table A2.1: Agriculture, Natural Resources, and Rural Development Sector—Success Rating of Completed Projects, 2008 Onward

Project	Loan/Grant Amount Disbursed ($ million[a])	Closing Date	Subsector	Rating	Lessons Learned
Tonle Sap Lowlands Rural Development Project	21.8	2017	Rural development	Successful	The project addressed aspects of agriculture, on-farm and off-farm productivity, and food security, both directly and indirectly, through different activities that focused on investments, training, and capacity building. The activities complement other projects in the project area funded by ADB and other development partners.
Northwest Irrigation Sector Project	26.9	2014	Irrigation	Partly successful	At all levels of involvement (executing agency, service providers, contractors, FWUC members), capacity constraints affected project performance. In these circumstances, the project design was both overambitious (e.g., by not focusing on fewer subprojects from the outset) and not rigorous enough in its analysis of the risks and assumptions inherent in many operations.
Tonle Sap Sustainable Livelihoods Project	20.0	2012	Rural development	Partly successful	The project demonstrates that, though institutionally complex, it is possible to improve natural resource-based livelihoods in a decentralized context. A more in-depth institutional analysis at all levels could have overcome some of the problems encountered during

continued on next page

Table A2.1 *continued*

Project	Loan/Grant Amount Disbursed ($ million[a])	Closing Date	Subsector	Rating	Lessons Learned
					implementation. Such analysis should be made an essential project design requirement for future development interventions of a similar nature in Tonle Sap.
Agriculture Sector Development Program	27.0	2011	Agriculture	Successful	The project delivered appropriate training to over 50,000 poor and nonpoor households in 18 districts of the four selected provinces. Shortcomings in the training of agro-enterprises and MAFF staff were not significant in the overall context of the project.
Stung Chinit Irrigation and Rural Infrastructure Project	25.6	2009	Irrigation and rural access	Partly successful	ADB and Cambodia's agencies did not fully appreciate the economic and technical implications of investing in irrigation at Stung Chinit. ADB's TA-based project preparation and loan processing cycle could not handle the uncertainties and complexities besetting the project; either a more intensive project design phase should have been required prior to loan approval, or a more flexible project implementation should have been designed.
Northwestern Rural Development Project	32.5	2009	Rural development	Successful	The consultative and participatory methods used under the integrated rural accessibility planning and in community development have provided villages and commune councils with hands-on experience in infrastructure prioritization and implementation.

ADB = Asian Development Bank; FWUC = farmer water user community; MAFF = Ministry of Agriculture, Forestry and Fisheries; TA = technical assistance.

[a] Including cofinancing.

Source: ADB.

Table A2.2: Agriculture, Natural Resources, and Rural Development Sector—Success Rating of Completed Technical Assistance Projects, 2008 Onward

Project	TA Amount Disbursed ($ million[a])	Closing Date	Subsector	Rating	Lessons Learned
Harnessing Climate Change Mitigation Initiatives to Benefit Women	2.8	2018	Rural development	Successful	Pilot projects enabled the development of women-led businesses within clean energy supply chains and demonstrated an alternative narrative about women's contribution to climate mitigation.
Innovative Financing for Agricultural and Food Value Chains	1.4	2017	Agriculture	Successful	A major lesson learned is that agribusiness value-chain financing models are highly diverse, dynamic, and context-specific, and that no single best practice can be found for simple replication in the region. The TA identified different sets of legal, institutional, and physical constraints that need to be addressed in each DMC to support identified innovative financing models.
Capacity Building for the Efficient Utilization of Biomass for Bioenergy and Food Security in the GMS	3.5	2016	Rural development	Successful	Significant contributions were made in strengthening institutional links and regional policy dialogue, harmonizing standards and certification systems, preparation of road maps in rolling out bioenergy standards, facilitating exchange of information and skills within the region, commercial testing and development of pilots as investment models, and capacity building.

continued on next page

Table A2.2 *continued*

Project	TA Amount Disbursed ($ million[a])	Closing Date	Subsector	Rating	Lessons Learned
Expansion of Subregional Cooperation in Agriculture in the GMS	1.1	2011	Agriculture	Successful	As the agriculture sector is large and important for all GMS countries, there is a tendency to be overambitious in formulating regional cooperation frameworks.
Implementation of the Action Plan for Gender Mainstreaming in the Agriculture Sector	0.3	2008	Agriculture	Successful	The key to the success of the TA project was the strong sense of ownership and commitment of the MAFF, particularly its government unit staff. All policies, strategies, and plans were developed by senior management at the MAFF, with the rigorous participation of key stakeholders.

DMC = developing member country; GMS = Greater Mekong Subregion; MAFF = Ministry of Agriculture, Forestry and Fisheries; TA = technical assistance.

[a] Including cofinancing.

Source: Asian Development Bank.

Table A2.3: Agriculture, Natural Resources, and Rural Development Sector—Assistance of ADB and Other Major Development Partners

Development Partner	Project Name	Period	Amount (million)
Agricultural Production			
ACIAR	Building a Resilient Mango Industry in Cambodia and Australia through Improved Production and Supply Chain Practices	2013–ongoing	$1.1
ACIAR	Improved Irrigation Water Management to Increase Rice Productivity in Cambodia	2014–2015	$1.1
ADB	Uplands Irrigation and Water Resources Management Sector Project	2015–ongoing	$60.0
ADB	Cambodia Agricultural Value Chain Infrastructure Improvement Project	2018–2021	$2.0
ADB	Southeast Asia Agriculture, Natural Resources and Rural Development Facility	2018–2023	$8.0
ADB	GMS Sustainable Agriculture and Food Security Program	2019–2025	$5.5
ADB	Agricultural Value Chain Competitiveness and Safety Enhancement Project	2021–2027	$110.1
ADB	Southeast Asia Agriculture, Natural Resources, and Rural Development Facility, Phase II	2020–2025	$6.1
EU	Promoting Climate-Resilient Livelihoods for Small-Scale Farmers in the Most Vulnerable Dry Land Areas	2011–2015	€1.8
EU	Improving Food Security and Market Linkages for Smallholders	2011–2015	€4.0
IFAD	Tonle Sap Poverty Reduction and Smallholder Development Project	2009–2023	$121.3
IFAD	Accelerating Inclusive Markets for Smallholders	2016–2023	$61.6
USAID	Micro, Small and Medium Enterprises Strengthening 1 and 2 Project	2005–2012	$26.5
USAID	Helping Address Rural Vulnerabilities and Ecosystem Stability	2010–2015	$56.8
World Bank	Mekong Integrated Water Resources Management Project, Phase III	2016–ongoing	$16.5
World Bank	Cambodia Agricultural Sector Diversification Project	2019–2025	$101.7

continued on next page

Table A2.3 *continued*

Development Partner	Project Name	Period	Amount (million)
Agro-Industry, Marketing, and Trade			
ADB	Implementing the GMS Core Agriculture Support Programme, Phase II	2012–ongoing	$19.3
ADB	Climate-Resilient Rice Commercialization Sector Development Program	2013–ongoing	$79.1
Australian Aid	Cambodia Agricultural Value Chain Program	2010–2015	$61.3
IFAD	Sustainable Assets for Agriculture Markets, Business and Trade Project	2019–2025	$124.5
World Bank	Agribusiness Access to Finance Project	2010–2014	$5.0
World Bank	Agricultural Finance Support Facility Cambodia	2013–2015	$0.7
Agricultural Policy, Institutional and Capacity Development			
ADB	Water Resources Management (Sector) Project	2006–2009	$1.9
ADB	Tonle Sap Lowlands Rural Development Project	2007–2015	$20.0
ADB	Tonle Sap Technology Demonstrations for Productivity Enhancement	2008–2013	$3.5
ADB	Capacity Building for the Efficient Utilization of Biomass for Bioenergy and Food Security in the GMS	2011–2015	$4.0
ADB	Support for Public–Private Partnerships	2011–2014	$0.3
ADB	Mainstreaming Climate Resilience into Development Planning	2012–ongoing	$11.0
ADB	Strengthening Coordination for Management of Disasters	2014–ongoing	$2.0
EU	Cambodia Climate Change Alliance	2009–2014	€2.2
IFAD	Project for Agricultural Development and Economic Empowerment	2013–ongoing	$43.2
IFAD	Agriculture Services Program for Innovation, Resilience and Extension	2014–2022	$86.6
JICA	Project for Establishing Business-Oriented Agricultural Cooperative Models (Cooperative Support and the Food Value Chain)	2013–ongoing	$3.8
World Bank	Strategic Program for Climate Resilience, Phase 1	2010–ongoing	$1.5
World Bank	Land Allocation for Social and Economic Development Project III	2020–2026	$107.0

continued on next page

Table A2.3 *continued*

Development Partner	Project Name	Period	Amount (million)
Water and Irrigation			
ADB	GMS Flood and Drought Mitigation and Management Sector Project	2012–2017	$45.0
ADB	Irrigated Agriculture Improvement Project	2019–2025	$119.2
ADB/AFD	Northwest Irrigation Sector Project	2005–2014	$28.4
ADB/AFD	Stung Chinit Irrigation and Rural Infrastructure Project	2000–2008	$18.2
ADB/DFAT	Water Resources Management Sector Development Program	2010–2016	$51.7
AFD	Rehabilitation of Prey Nup Polders	1997–2008	$113.5
AFD	Water Resources Management and Agricultural Transition for Cambodia, Phase I	2014–2019	$12.9
AFD	Water Resources Management and Agricultural Transition for Cambodia, Phase II	2017–2025	$55.0
DFAT	Cambodia Agricultural Value Chain Program, Phase 1	2010–2015	$57.9
DFAT	Cambodia Agricultural Value Chain Program, Phase 2	2016–2021	$86.8
Government of Japan	Approximately 22 small-scale rehabilitation projects	2009–2016	$2.0
JICA	Technical Services Centre for Irrigation Systems, Phases 1 and 2	2001–2009	$5.0
JICA	River Basin Water Resources Utilization in Cambodia	2014–2019	$4.3
JICA	Southwest Phnom Penh Irrigation and Drainage Rehabilitation and Improvement Project	2014–2026	$58.0
JICA	West Tonle Sap Irrigation and Drainage Rehabilitation and Improvement Project	2011–2016	$44.0
KOICA	Dauntri Dam Development Project	2014–2019	$46.7
KOICA	Krang Ponley River Basin Development (multidams)	2006–2012	$26.0
KOICA	Mongkol Borey Dam Project	2011–2015	$18.0
KOICA	Sala Ta Orn Irrigation	2013–2017	$36.0
PRC	Kanghot Irrigation Development, Phase 1	2010–2015	$49.9
PRC	Kanghot Irrigation Development, Phase 2	2012–2016	$33.9
PRC	Multipurpose Dam Development	2012–2017	$99.2

continued on next page

Table A2.3 *continued*

Development Partner	Project Name	Period	Amount (million)
PRC	Prek Stung Keo Water Resources Development	2011–2015	$47.0
PRC	Stung Chikreng Water Resources Development, Phase 1	2014–2019	$45.0
PRC	Design and Build of Stung Pursat Dam	2011–2015	$63.0
PRC	Stung Sreng Water Resources Development 1	2011–2015	$52.0
PRC	Stung Sreng Water Resources Development 2	2014–2017	$45.0
PRC	Vaico Irrigation Development, Phase 1	2012–2017	$99.0
World Bank	Mekong Integrated Water Resources Management Project	2017–2021	$15.0
World Bank	Water Supply and Sanitation Improvement Project	2019–2024	$57.5
World Bank	Sustainable Landscape and Ecotourism Project	2019–2025	$53.2

ACIAR = Australian Centre for International Agricultural Research, ADB = Asian Development Bank, AFD = Agence Française de Développement, DFAT = Department of Foreign Affairs and Trade (Australia), EU = European Union, GMS = Greater Mekong Subregion, IFAD = International Fund for Agricultural Development, JICA = Japan International Cooperation Agency, KOICA = Korea International Cooperation Agency, PRC = People's Republic of China, USAID = United States Agency for International Development.
Source: ADB.

www.ingramcontent.com/pod-product-compliance
Lightning Source LLC
Chambersburg PA
CBHW050051220326
41599CB00045B/7366